RICHARD ROBINSON is the author of seven books of popular science including the Science Magic series (C   , Press), which was short-listed Aventis Prize. He works full-time as a science presenter, and regularly performs demonstrations around the world from Boston to Beijing. He was also a founder member of *Spitting Image*.

WHY THE TOAST ALWAYS LANDS BUTTER SIDE DOWN

The Science Of Murphy's Law

Richard Robinson

ROBINSON
London

To Dad, better late than never

Constable & Robinson Ltd
3 The Lanchesters
162 Fulham Palace Road
London W6 9ER
www.constablerobinson.com

This edition published by Robinson,
an imprint of Constable & Robinson Ltd, 2005

A copy of the British Library Cataloguing in
Publication data is available from the British Library

ISBN 1-84529-124-7
ISBN 978-1-84529-124-2

Printed and bound in the EU

5 7 9 10 8 6 4

Contents

Acknowledgements

Contributors to this book have been nearly everyone I have met, and most inanimate objects I have ever come across. In particular, Rory Fidgeon and Sam Hutton for checking it through, and Bill Parish for the maths. Thirzie, Philip, Judy and Leonie have at some time or another had to cope with me while I struggled. Much respect to them.

Preface: Murphy's Law

The law of laws, Murphy's Law, sits above all others like an uninvited guest at the feast. Whatever your field of endeavour Murphy is there to trip you up, delay and frustrate you. The more your endeavour, the greater the uptrip.

WHATEVER CAN GO WRONG, WILL GO WRONG

Like all great laws it is awesomely simple in its formulation. Like all great laws it is, once uttered, a self-evident truth. The longer you live with it, the more you realize what a vice-like grip it has on the planet. You can't dodge it. You can't put your butter on the other side of the toast. The Law and its by-laws hedge you in on all sides:

WHATEVER CAN'T GO WRONG WILL GO WRONG

TRYING TO MAKE THINGS BETTER ONLY MAKES THINGS WORSE

ANY ATTEMPT TO DO NOTHING, SO NOTHING CAN GO WRONG, WILL GO WRONG

Originally discovered by technicians working on crash tests at Edwards Air Force Base in California, USA in 1949, Murphy's Laws spread quickly. They were spotted everywhere: people suddenly noticed the way that buses always came in threes, desperately needed objects would become invisible, vital components would, if dropped, roll under the heaviest cabinet. The list lengthened. Not just objects, but animals and people entered the lists: Nothing works when people watch? As soon as your hands are tied your nose itches? It happens to us all.

Is there a rational explanation? That's what this book is here to provide. In fact, a closer look at Murphy's Laws can give us some insights into us and the increasingly tangled lives we lead.

MURPHY IN THE MODERN WORLD

Murphy's Law is the child of the modern world, with all its mind-warping complexities. If you want to find a time of pre-Murphy innocence you need to look way back to the late Stone Age, about 5000 BC. To get an idea what life was like then, look out of the window and in your imagination sweep away the houses and gardens, cars and roads. Keep sweeping until you're down to bedrock. Next remove nearly all the people. Plant trees. Lots of them. Add a sprinkling of goats. Now you are back in Palaeolithic times. Everything our ancestors possessed was made from bits of tree, bits of stone or bits of goat. There's not a lot you can make with those kinds of raw materials; and there's not much can go wrong either. So my great- great- great-great- . . . great-grandparents very rarely came across Mr Murphy. And the world had been the same for their parents and grandparents way back to their distant ancestor, the lemur. Lemurs, monkeys, apes and humans evolved over millions of years to be happy in this simple, unchanging, pleasantly unpretentious landscape.

Compare that with the cacophony of today. In just a couple of hundred years modern humans have drilled beneath the simple prehistoric hills to haul out 92 interesting elements, which can be arranged to form 9,292 kinds of exotic molecule, from which 9,292 x 9,292 wonderful gizmos can be built, which can cause 9,292 x 9,292 x 9,292 maddening problems. To house the gizmos modern humans have spread housing estates, warehouses and leisure centres over the Palaeolithic hunting grounds. In 2000 BC the average family of ten could live, with all their tools and toys, in a space no bigger than a modern kitchen. Nowadays people in the West often buy second homes simply to house a mass of knick-knacks they claim they can't possibly live without. Likewise the simple tribal systems of 4,000 years ago have been swamped by the rich, complex social weavings within our vast cities. The social weavings are the consequence of the gigantic rise in population. The entire human population of the Stone Age world could fit into one medium-sized modern town.

We have come a long way since the Stone Age, but we haven't brought our brains with us. Our minds still dwell in Palaeolithic dreams. Where the Palaeolithic mind meets the modern world, there sits Murphy. How do our Stone Age minds react to the plastic age? By getting completely fazed by it. As anyone who has watched an adult wrestle with a childproof lid will testify, each forward step of progress leads us on to another banana skin of Murphy.

This book has been written to answer two questions: Why do inanimate objects do what they do? And why do humans get so peeved about it?

The latter end of the book is to do with inanimate objects. Shouting at inanimate objects takes up, in my experience, far too much of the day. But it can be very calming, when the laundry comes out of the washing machine once again neatly tucked inside the duvet cover, to know that there is actually a rational explanation. (In fact the clothes inside the duvet cover can lead us to greater insights about the Universe; why rubbish on the streets lies where it lies, why the Sargasso Sea looks like it looks and why townies move to the countryside.) In Chapter 8 there are 80 sane explanations of baffling but common phenomena. But this begs the question; if there is a scientific explanation, why are we still puzzled by everything?

The answer lies within ourselves, for very often Murphy is in the eye of the beholder. Our minds see the world conspiring against us, when the world is entirely innocent. So the first part of the book looks at how our brains help us to get everything so dreadfully wrong.

THE JIGSAW

Understanding the world is like doing a jigsaw; we look at the pieces, check them against the guide picture on the box, then try to fit them together. For our mental jigsaw the 'pieces' are the messages our senses send to the brain. The 'guide picture' is the brain's expectations and memories used to analyse the messages. Fitting them together is a task for the whole brain, and all three stages are suffused with Murphy's malign influence.

Chapter 1 discusses how the senses pick up the jigsaw pieces. As jigsaws go, this sensory jigsaw is a pretty tough one. The picture changes ten times every second (that's how often the brain updates its view of the world). Secondly, there are a million jigsaw pieces; a million nerve impulses surging towards our skull every tenth of a second. It's a tidal wave of jigsaw pieces, and we have to make sense of them.

The best we can do is grab at a few of them as they rush past, and try to guess the rest of the picture from what we've got.

In chapter 1 we will see how easy it is for us to get it wrong. We will discover some of the techniques we use to filter the input, such as habituation and attention mechanisms.

The illusions in Chapters 1 and 2 are deliberately designed to send the wrong messages, to trip up the senses. In a laboratory (or in this book) we can laugh them off because we know they're coming, but out in the street where we don't expect them they can floor us (quite literally in some cases; I speak as one of millions who have walked into a glass door that shouldn't have been there). Many Murphy's Laws stem from our simple inability to see what's right in front of our noses.

Once having worked out what jigsaw pieces we've got, and measured them (in **Chapter 2**), our brain has to work out what they mean. In **Chapter 3** we examine the role of the 'guide picture'; our memories. In our mental jigsaw the sights and smells of daily life give us patches and dots which send us reaching for the memory banks. Is that dot an eye? Is that blue patch a bit of sky? Our memory gives us an idea of what to expect. But when Murphy's Law is up and active we ignore the facts and believe the expectation: late at night the shadow in the corner of the bedroom can remind us of a human outline. Is it a burglar? Murphy's Law says that the more we need sleep, the more that shape in the corner will turn into a burglar, probably armed with a giant knife. Sometimes our memories fail completely – we enter a room to look for something, then wander round like idiots because we can't remember why we're there. (Older readers who thought this was a sign of creeping senility will be pleased to find it happens to ten-year-olds as well.)

The next step involves making connections. In **Chapter 4**, the pieces of jigsaw are put together. Here the chances of Murphy's Law striking us down increase logarithmically, because we are criminally bad at putting our puzzle together correctly. At this point it becomes clear there is a big difference between our mental jigsaw and a wooden one. Wooden ones only fit correctly one way; the brain jigsaw seems to be infinitely bendy, as if made of jelly. So if two facts don't seem to fit together, they can easily be twisted and punched until they do. The compulsion to hammer together pieces that don't fit causes us big problems. For instance, if you are the bearer of bad news, be afraid. The tendency of tyrants to shoot the messenger is no myth. They fit together 'bearer' and 'bad news' until satisfied that one is responsible for the other, then shoot him.

Although we think we've finally put together our puzzle, it's not over yet. The jigsaw and all around it is flowing with a thick, seething, electric sauce – the emotions. **Chapter 5** studies what a difference emotions make. This is not just about the grand passions – love, hate, fear – but about the whole spectrum, from the first yearning kiss to the 99th yawn. Nothing, but nothing, is free from emotions, whether it's a python or a garden hose. Research with modern brain-scanning techniques has demonstrated that when our eyes first light on something, the first thing we do (and it happens so quickly and so unconsciously that we are not aware of it) is check if we should run away from it. It seems absurd to imagine us, with all that expensive education behind us, looking at a garden hose and wondering whether to flee in terror. But hoses and snakes look quite similar at first glance, and if this long thin shape turned out to be a python we'd off and away literally before we are aware of what we're running away from, and grateful for the reflex that made us do it. (To see what a horse does when it sees a hose, turn to p. 87.) The part of the brain responsible for this reflex is the amygdala, the seat of our emotions. The amygdala rules over all else like an Egyptian Pharaoh. The state of our passions governs how we see what we see, and sometimes makes us see things strangely. – 'To a man with a hammer', the saying goes, 'everything looks like a nail'. It works for humans and it works for animals – horses bolt when they see garden hoses because their emotions have told them to run before their intellect has time to tell them it's not a python.

Chapter 6 is the social one. Many of Murphy's Laws apply when we are in company. It is the despair of psychological researchers that people are completely different in groups to how they are alone. However much you probe and calibrate them when they're on their own, you can bin your graphs and tables when they get together with their mates. Thoroughly domesticated human beings will go to a football match or off for a weekend paint-balling with a group and be transformed into a screaming, murderous, eyeball-popping life form unrecognizable to their families.

People can be made to believe the most extraordinary things when they are caught up in a crowd – this of course is the bedrock of politics: 'Some of my best friends are

German' gives way to 'Kill the Square-heads!' Arguably our success as a species stems from an ability to lock our individuality in a bottom drawer for the time it takes to help our party or army or tribe or corporation to defeat the enemy. In peacetime, however, there can be a serious lack of focus amid a cluster of humans. Anybody who has wit- nessed the workings of a committee will vouch for the strangeness of some forms of group behaviour. So the sixth area where Murphy is found is in committees, crowds, parties and family life.

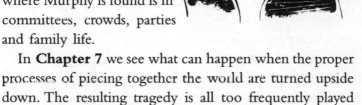

In **Chapter 7** we see what can happen when the proper processes of piecing together the world are turned upside down. The resulting tragedy is all too frequently played out in different forms around the world.

Finally, in **Chapter 8** we look at the inanimate world, and discover how much animation we impose on it. I hope you will find answers to some of life's mysteries here. You may find answers where you didn't know there were questions. And I expect you will think of many questions for which the book provides no answers: Murphy's Law will ensure that I think of several very important things I forgot to mention the day after the book is published.

Enjoy.

Note on the Shape of the Brain

The pictures overleaf outline which parts of the brain do what, but they come with a warning; they are very simple, which means they are undoubtedly inaccurate. Reading this book should alert you that anything which is easy to understand is bound to be wrong, so take the guide below with a pinch of salt. For instance, the amygdala is not the only area that deals with emotions, but it does seem to be central to that complex chain of processes, so it is given that simple role in this book. You should not try to build your own brain based simply on the design below. With approximately 100,000,000,000 neurons in the brain, each one with dozens of connections, you don't need to have a big brain to realize there are more connections possible than there are sub-atomic particles in the universe. So understanding how it all works is never going to be possible. If the brain were simple enough for us to understand it, we'd be so simple we wouldn't be able to.

Amygdala
The emotional centre of the brain. Whatever the intellect may say, it is the amygdala which provides the emotional flavour of our experience. Without this, the world is flat and grey. With it, fighting, fleeing, loving and hating are possible.

Hippocampus

Memories start here, created out of inputs from the cortex, the various sense organs and the amygdala. Once built, memories are stored around the cortex. In cases where the hippocampus has been damaged, memories prior to the accident remain intact, but later events leave no memory trace at all.

Hypothalamus

The main function of the hypothalamus is *homeostasis*, or maintaining the body's status quo. It monitors the blood for sugar, temperature, pressure, oxygen content, and receives information from all zones of the brain. It takes action on what it finds in two ways — by sending signals to various areas of the brain to change behaviour, for instance if the oxygen levels are too low it causes the body to struggle to find air, instigating a 'panic attack'; and by chemical signals, for which it employs the pituitary.

Pituitary

Sometimes called the 'master' gland of the endocrine system, because it controls the functions of the other endocrine glands. The pituitary gland is no larger than a pea, and is located at the base of the brain. It secretes a variety of crucial hormones on the command of the hypothalamus to stimulate the adrenal glands, the thyroid gland, the ovaries and testes.

Thalamus

The fact that so little is known about this central, and rather large body, when so much is known about, for instance, the tiny hypothalamus, indicates just how much there is still to discover about the human mind. All we know is that nearly everything makes regular contact with it. It seems to be a 'last pit stop' for information going to the cortex – sorting and sifting signals before sending them forwards or backwards.

THALAMUS

Putamen

A particular kind of memory has its base here. The learning of technical skills, involving coordination of limbs, eyes, memories and dealing with gravity, etc. So playing the piano, skateboarding, playing the piano on a skateboard, all these are coordinated by the putamen.

PUTAMEN

The cortex

Compared to the complicated web of jobs in the lower regions, the outermost layer of 'grey matter', about 1 cm thick, seems quite straight forward. Each area has its own clear boundary, although the intricate detail of what goes on inside the bound-ary can only be broadly described.

Cerebellum

Deals with balance and muscle coordination. How impor-tant this is can be seen when it malfunctions. Problems include shaking limbs, roaming eyes and lack of focus, uncontrolled blinking, chewing and swallowing difficul-ties. Even the task of keeping still requires constant control from the cerebellum.

Broca's area and Wernicke's area

Dealing with language involves two processes – Wernicke's area copes with understanding words; Broca's area coor-dinates speech production.

Visual cortex

The visual cortex deals with input from the eyes. Analysing it for colour in one area, shape in another, movement, depth, and so on. The more complex judgement of 'significance' happen towards the periphery, linking to memories, and relating to the body as a whole.

Sensory cortex
The body is mapped out here in regular order, from top to toe, starting with the head at the outer edges and ending with the toes up at the top of the head. Not only are the touch sensors monitored here but also positions and angles of the various limbs, so the area is usually called the 'somato-sensory' cortex.

Motor cortex
Deals with instructing the muscles to do the mind's bidding.

Pre-motor cortex
Calculating which muscles must do what, and how much, and for how long, is done here. The pre-motor cortex plans actions on the basis of instructions from the areas just in front.

Frontal and prefrontal lobes
This area has received a lot of attention from psychologists, because it is so important, yet so much more amorphous than the areas behind. Here, it seems, sits will-power, judgement, ethics, conscience, mediation and inspiration. Trying to analyse what exactly is meant by all those terms, trying to find a particular weave of neurons relating to, say, 'respect for one's elders', is either the holy grail of neuroscience or a will 'o the wisp, depending on the construction of your frontal and prefrontal lobes.

1

Taking It All In: The Senses

The first step to doing a wooden jigsaw is to empty the box on to the table, turn the pieces over and see what we've got.

To do our mental jigsaw, too, we should look at the pieces. But it's a very big jigsaw. Very big indeed. Our brain is dealing with a huge flood of information coming from millions of nerve endings in our eyes, nose, tongue, ears and skin. If this were a jigsaw, there would be millions of pieces to it. And since the mind updates the picture every tenth of a second, the pieces of mental jigsaw have to change ten times a second to match. How does the brain cope with this? Quite simply it ignores nearly all the incoming signals, grabs a few, and fills in the remaining spaces by guesswork.

You'll be amazed how much guessing goes on. This will give you an idea: cover the bottom half of this page as far as the jigsaw piece below. Look at the eye in the middle of the piece. Don't look at anything else, but uncover the bottom of the book. Now try to read the words below the jigsaw piece without moving your eyes.

At any given moment your eyes can only truly know about an area the size of that jigsaw piece. The rest of the picture is rumour and hearsay; you are guessing it.

Later on we will look at how the guessing is done. In Palaeolithic times, in a world made up of rocks, trees and goats, our ancestors could guess pretty accurately what the signals meant; nowadays much less so. New colours, smells and sounds overwhelm our senses, and Murphy stalks the brain.

POTENTIATION

Let's look through the senses and see how easy it is to get it wrong.

HEARING

Nothing could be simpler than sounds – waves of compressed air. It doesn't matter how many different noises, speeches, tunes, hisses or burps there are in the room around you, what goes into the ear is simply waves of compressed air. The eardrum vibrates, the vibrations bounce through to the cochlea, which translates the vibrations into nerve impulses, which travel to the auditory cortex. Then the fun begins. With fantastic forensic skills the auditory cortex separates the raw signals out into individual sounds. Palaeolithic hunters could separate birdsong, rustling trees and a bubbling river from the tiny sound which they wanted to hear, a deer munching on leaves behind the next bush. Nowadays we are surrounded by a different din. In a party, recorded music replaces the birdsong, clothes do the rustling, a hundred people's burbling replaces the river, and the particular sound we want to listen to is not a deer munching on leaves but a guest nibbling on a snack and telling us about his hobby.

Although he is speaking very quietly, and there's a much louder din all around, we are able to hear him amazingly clearly. This phenomenon is called the 'Cocktail Party Effect' because the aptly named Colin Cherry, who first investigated it, found it particularly noticeable at cocktail parties[1]. His pioneering work in the 1950s was not about helping us to understand cocktail parties, but to help air-traffic controllers dealing with the confused babble of voices in their headphones. Cherry's work with air-traffic controllers showed how we all select what we want to listen to, and manage to ignore the rest. What it demonstrates is that as soon as the brain has separated the sounds and assessed them, they are practically all dumped, leaving only the sounds from that guest and his hobby.

Actually it's not quite that simple, as Murphy points out every now and then. The other noises are not excluded instantly; they are being unconsciously monitored. Proof of this is:

If you say the word 'sex', the room goes very quiet

Suddenly everyone at the party is listening to you. How could they have heard that word above the din? How embarrassing. This Murphy's Law shows up a remarkable ability of the auditory cortex. The babble that the brain seems to be ignoring is actually being unconsciously checked for key words before being discarded. Any sounds the brain considers interesting are being pushed through to consciousness. 'Sex', unsurprisingly, is considered a very interesting word.

Other significant words or sounds can emerge from the din. A mobile phone with the same ring-tone as yours will make you look up. If someone says your name your ears will burn – and it has been noted that if you do hear your name you also hear several of the words that preceded it; proof that the brain holds the outside noise in store for a few seconds.

A confidential whisper is heard better than a loud shout

The brain monitors the surrounding noises thoroughly. Subtle changes to the sound envelope are considered worthy of special attention, even if they involve a lowering of volume. If the couple next to you drop their voices to a whisper you are alerted; your ears strain to hear what they are saying.

As you pry on the neighbours you don't hear all their words, only parts. The brain adds in what it thinks the missing bits are. Generally you do a pretty good job of it, but sometimes you can get things wrong. Here's an analogy which shows how. Look at the words on the opposite page. You are given part of them, you make up the rest. Your brain reaches for what it guesses must be the right words, but of course it's got it wrong (turn the page opposite to find out how wrong).

The brain tends to hear what it wants to hear from the fragments it receives.

Three drunks are rolling down the road one morning. 'Sunny today isn't it!' says the first. 'No it's not, it's Thursday,' says the second. 'So am I,' says the third, 'Let's have another drink'.

I had a Murphy-chat with a Chinese woman in Beijing recently. Her English pronunciation was not perfect, but I thought we were doing fine. She had the job, she said, of finding horses for people coming to live in Beijing. I mentioned that I hadn't seen any horses since I had arrived.

MERRY CHRISTMAS

'Oh yes', she said, 'everyone in Beijing has a horse'. 'Everyone?' I spluttered. 'But I haven't seen a single one'. 'Where else would they live?' she said. We paused, and it dawned on me that what we were talking about was 'houses'. A small difference in pronunciation resulted in a strangely surreal conversation.

PROPRIOCEPTION

Proprioception is the Cinderella sense, always excluded from the company of the Big Five (hearing, seeing, smell, taste and touch). But when you sleep wrong and wake up with a dead arm dangling uselessly off your elbow, you should be grateful that generally speaking you know where your limbs are. That is proprioception. The ability to walk, run and stay upright depends on this knowledge, and to that end feedback messages are sent from all over the body to the cerebellum. (The name is Latin for 'little brain'. It really does look like a spare brain, hanging just underneath the main one at the back behind the brain stem.)

Your numb arm recovers soon enough, but in *The Man Who Mistook his Wife for a Hat* Oliver Sacks describes how one of his patients woke to find somebody else's leg in the bed beside him[2]. In terror he threw the horrifying thing out of the bed; but because it was actually his own leg, he too was flung out and they all landed on the floor together. You don't think about it, but proprioception is a vital sense. That patient had lost it. His limb was perfectly normal in every other way, but it was not sending out any messages about

MFBPX GIJBISJMAS

its location, so it had simply stopped being part of him. What is of deep interest to us here is that the patient's mind had to be able to explain what was happening. So it was that he found himself believing an impossible belief, that this leg which clearly looked like his, and was in the right place, was actually someone else's.

You can fool your proprioception by causing muscles to go into spasm. Stand next to a wall, as in the picture. Keeping the arm straight, try to force it through the wall. Keep straining outwards like that for 30 seconds, then relax. When you look down you will find your arm in the wrong place, floating above where you think it ought to be.

Standing on a cliff edge is the one time you need to be cool, and the one time you can't be

(Why the World is Wonky)

At the cliff edge the normally smooth operation of the cerebellum is set aside in favour of another thing the brain is very good at – panic. We have an inbuilt fear of empty space below us. It shows in babies as soon as they become mobile; if they are crawling towards their mothers and a hole appears below them, even if it is obviously covered with thick glass which they could crawl over, they will go no farther forward.

As you stand on the edge of your cliff your depth-fear response kicks in and your cerebellum is rendered helpless. The basic fear response is to freeze. This policy clashes with the cerebellum's job of fine movements to keep you upright. The result is a tendency to sit down firmly and

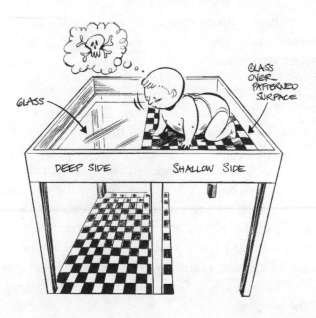

refuse to budge. Lots of practice at walking along cliffs teaches you to trust your cerebellum and let it get on with its job.

When you start to learn a new skill, like acrobatics, your cerebellum has to open a whole new book of calculations related to what happens when you are upside down. For me, being upside down is as familiar as being on another planet; for circus acrobats who do it for a living, walking on their hands is no more difficult than walking on their feet. Over the years their cerebellum has done all the calculations, and now it comes naturally.

Children instinctively give their cerebellums practice at balancing by climbing trees, walking along walls, playing on climbing frames, etc. But the opportunities to practise are limited by the spread of concrete. Falling out of a tree is educational if you land on grass, but life-threatening if you land on concrete. Rather than provide the grass, the guardians of our safety tend to remove the trees. Now our children have nothing to practise on. Less practice means less confidence, which means our children become less safe, not more so.

SEEING

With the eye, as with everything, most of our 'observing' is make-believe. The structure of the eye shows why this is. There are 130 million light-sensitive retina cells at the back of the eye, but only 1 million cells in the optic nerve, which takes the messages to the visual cortex at the back of the head. So 129/130ths of the information the eye receives is either thrown away or heavily compressed (like a jpeg, but much more so). The visual cortex uses its sophisticated software to work out the picture. Usually it's right. But Murphy's Laws are all about it being wrong; about us saying, 'Aah, look at the way that cute dog is smiling at me' just before it bites our hand.

You can observe a lot, just by watching.

(One of the famed sayings of Yogi Berra, the baseball player and manager, originator of the cryptic, 'The game ain't over till the fat lady sings', and inspiration for Yogi Bear.)

The illusion below seems to show two triangles, one behind the other.

In fact there's no such thing. There are three circles with slices missing and three Vs. Your brain has taken those pieces of visual jigsaw, used its imagination and is now inventing a triangle running between the circles, and then another imaginary triangle behind the first imaginary triangle. And more: to make it clearer still, your brain is also adding colour; the top triangle seems to be a lighter white than the background. So what you see is a mixture of what is there and what you have added.

TOUCH

Millions of skin receptors all over your body send messages to your brain. The messages are not only about touch, but also pain, pressure, heat and cold. Mostly you don't bother with them. But you can clock on to them easily: You can feel your shoes gripping your feet right now . . . Now you can forget them again.

As soon as you lie down to sunbathe on the grass, all the insects in the world start to crawl over you

Flinging your clothes off and flopping on the grass is not as easy as it sounds. There are creepy-crawlies down there in the soil. They might be hungry, and you might be lunch. You feel them starting to walk across your skin. You look down – they've gone. You shut your eyes again – they return with their families. They're going to feast on you. But it's all a phantasm. Open your eyes and they've disappeared again. Why do you imagine creatures crawling over you?

Your touch sensors also send a lot of random noise to the brain – minor tics and tiny clicks. Normally you ignore them; they are just some of the trillions of messages the brain dumps. But sometimes you become hyper-aware of them. When you are scared that beetles, wasps, ants or starving millipedes will discover you lying exposed and tasty on the lawn, you listen to the tics and clicks and read 'possible hunter-killer' into all of them.

TASTE AND SMELL

The taste senses are wired up slightly differently from the other senses. They have a hot line to the amygdala, the brain's fear centre. Disgust is an immediate and powerful reaction to some tastes; correctly so, because without it we could poison ourselves. When your head jerks away from the smell of milk that has gone off, it's as automatic as if the milk had punched your nose.

'Good for you' means 'yucky'

Food-stuff manufacturers know exactly what we desire to eat most, and zoom in on it with unerring accuracy; sugar, fat and salt. Our mouths water at the smell because we are sniffing with a million-year-old brain, one that was designed for living in a barren plain with scant opportunity to eat anything but tree bark. The sight of a bee would set our forebears off on a long hunt, tracking the bee back to its hive, climbing the tree where it is, withstanding a thousand stings for the sake of the glorious and incredibly rare taste of sweetness. That instinct is still there and guides our groping hands towards sweetness wherever and whenever it becomes available. The modern world puts sweetness at our fingertips 24 hours a day, with dire consequences for our health. If we wait patiently our brains may in the future evolve a correct sense of proportion, an instinct for the balanced diet. But while we're waiting we can't help feeling peckish, so we'll have just one more ice cream . . .

The sixth sense

Is there a 'sixth' sense? Most of us have 'felt' someone stand-ing behind us at some point, but how did we know they were there? Perhaps the answer lies within the unconscious processing that we now know goes on beneath the conscious layers. In the same way that your ears monitor the surrounding hubbub in a cocktail party and pick up on the sound of your name, they monitor the tiny echoes and reverberations of an empty room and spot any change caused by a sound-absorbing shape approaching behind. Blind people learn to listen much harder than the rest of us to the sounds a room makes.

As we will see in the next section, we have to be looking before we can see, and listening before we can hear. A lot of what's around never gets past the entrance to the doors of perception, but there seem to be quite a lot of sensations that can slip quietly into the mind and hover in the corner unnervingly.

Conscience: the inner voice that warns us someone might be looking

(H.R. Mencken, 1880–1956; *Chrestomathy*, 1949)

The psychologist Richard Wiseman has investigated the paranormal enough to feel ready now to haunt a house to a predetermined schedule[3]. He has found the factors which make people feel spooked: sudden but subliminal changes in temperature around the legs; hidden magnets, which can have an effect on hairs on your skin; infrasound, which is too low to be audible, but affects the body. All these will affect the hauntees, but subliminally, so they are aware of a sense of unease, but can't think why. Their imaginations will do the rest.

When you think there's something you don't quite like about somebody, but you can't put your finger on it, you may be seeing something unconsciously that your conscious mind can't see. You can read more into a person's body language than you realize. And people give themselves away more than they know. Frame-by-frame analysis of videos of conversations can reveal 'micro-reactions', rapid gestures which flit across the face, revealing the true feelings beneath the mask, but only at a subconscious level.

ATTENTION

For many of Murphy's Laws, the problem arises not from the senses as such, but from how much we are attending to them. For Murphy, attention is 9/10ths of the law.

Attention is what enables you to listen to the conversation in front of you and ignore the surrounding racket at the 'cocktail party', to suddenly decide that twinges from the skin mean that insects are attacking you. If you don't ignore most of the million signals coming into the ears, eyes and skin, you'll suffer sensory overload. The ability to zoom in on one thing and eliminate the others is a marvel. It seems as though the brain turns a searchlight on the world, which picks out some things and leaves others, blessedly, in the gloom.

But what to ignore and what to attend to? In the booming, buzzing modern world a lot of Murphy's Laws show up the difficulty of telling the wood from the trees, in a world where woods and trees have been smothered in chrome and plastic.

The longer you look at the page, the more the words don't go in

There are two parts to reading, and on this occasion they are both peeling away from each other: The ability to scan a line of text from left to right, then jump to the beginning of the next line accurately, is learned young, locked away safely in the putamen (see p. xxi), and performed on demand. The ability to understand the words while doing this is another department. It is perfectly possible

A Bedouin folk tale shows how hard it can be to control one's attention mechanisms:

A Dervish taught a Bedouin how to make gold out of straw: 'Simply put the straw in boiling water', he said, 'and stir it for three hours. But remember, while stirring don't ever think of a pink elephant . . .'

(I am the research document) to finish reading a page and suddenly realize you have not the faintest idea what it was about. The reflex of scanning hid the fact that you weren't really attending.

Making sense of a sentence is more complicated than it seems. We think that a sentence gradually emerges as each word follows the last. But this cannot be true – it is impossible to know from the first few words what the meaning of the whole sentence will be. Those words have to be held in limbo until enough new words are added to make

a sensible sentence. The holding of

words in

limbo while you wait for the sense to

emerge is a

pretty smart

act.

For words, as for sentences, understanding comes in globs. Aoccdrnig to rscheearch at Cmabrigde Uinervtisy, it deosn't mttaer in waht oredr the ltteers in a wrod are, the olny iprmoatnt tihng is taht the frist and lsat ltteer be at the rghit pclae. The rset can be a tatol mses and you can sitll raed it wouthit porbelm. Tihs is bcuseae the huamn mnid deos not raed ervey lteter by istlef, but the wrod as a wlohe.

It mkeas me wndoer wteehhr spttnog slplnig mtskaeis is at all esay.

Each solution opens the door to a more difficult problem

(Distilled from John Peers, Henry Kissinger, Krishnamurti, Stephen J. Gould)

By focusing attention on installing that new software package on the computer, you have avoided attending to many things. The task in front of you is not only vital, as you have patiently explained to everyone in the house, it also acts as a hedge between you and, for instance, the washing-up. Not because the washing-up is a problem, you understand, but because on the way to do it you'll pass the screwdriver that needs returning to the toolbox, which is next to the vase that needs mending with the glue, a new tube of which has to be bought from the hardware store, while perhaps at the same time doing a little shopping in the car, which should be taken in for a service, because tomorrow is the deadline for the road tax . . . and for the annual accounts, now you come to think about it, which you wish you hadn't. All things considered, it's best to just attend to the computer for now. One thing at a time.

For any given buffet there is one too many plates

Here is an assault course for the attention mechanisms. The buffet lunch gives a fine chance to play Double or

Drop over an expensive carpet. With each new plate, fork, serviette and glass your fingers have to cope with while you are trying to impress the new head of department, the probability rises that Murphy will show up as an unwelcome guest. You are teetering on the edge of overcapacity.

What you do now depends on three things, according to psychologists: **difficulty, practice** and **similarity**[4]. A lot of your attention has been allocated to being witty, because that's **difficult**. You shouldn't have to attend too much to the plate-balancing tasks because yesterday afternoon you **practised** holding a plate, glass, serviette horizontal with the same hand. But the tasks are unfortunately too **similar**, so when a vol-au-vent begins to leak over your chin as you squeeze it between your teeth, and you can't reach for the serviette because of the fork between your fingers, then you have reached the limit. This is when Murphy rings the phone in your inside pocket. The extra attention needed to judge whether to answer it, finish telling the joke or just stand there dribbling mushroom sauce finally tips you over the edge. Something is going to give. It's the plate which moves from the horizontal, a mini-pizza slips over the edge. As you grab for it the fork strikes the glass which spills, while everything else slips to the floor, apart from the vol-au-vent trickling down your lapel. The good news is that now you are unencumbered you are much better placed to attend to developing your witty story for the head of the department . . .

When your hands are tied your nose itches

There is only a limited amount you can focus on at any time. But other messages are being received unconsciously. Then, when the immediate task is completed, the mind has a chance to deal with other things. Your nose has been itching for some minutes, but only now does it enter consciousness.

The old image of the puzzled professor scratching his head stems from this switch of attention. While he is focused on the task in hand, he has thoughts only for that task. As soon as he takes a step back, other things can intrude, such as the itchy scalp.

There are many examples of what happens when the working memory is finally free — when the desktop is cleared. For instance . . .

You think of 10 important things to remember just as you are falling asleep

The last thing you want to haunt you as you drift off to sleep is the things-to-do-URGENT list. But now you've remembered them you know very well that if you leave them, those important things will be forgotten. You force yourself awake, crawl out of the bed, hunt for a pen and paper, scratch down the list and crawl back into bed. Then you lie staring at the ceiling for the next hour, unable to sleep. In the morning you find you've remembered everything anyway.

Before you slip into sleep, many of the day's events will take a lap of honour round the skull. This time they will not be running in sequence, but in parallel. They'll all be equally important. New links can be made. Here we can see the extent of the mind's unconscious processing. While you were dealing with the brain's crowded desktop today other things were being assimilated unconsciously. In the relaxed atmosphere of near–sleep, other important connections are made.

This Murphy's Law gives psychologists a clue to what dreams are about. While you are asleep, all the events of the day will be processed. Important events will be linked to other experiences in your life. Perhaps today you had your first trip on Nemesis, the most stomach–churning fairground ride in the Universe. Tonight it will be bedded in with other memories of being turned upside down, having your stomach lurch into your mouth, being surrounded by screaming, the fear, the pressure of unimaginable forces, the helplessness. You may have a dream about giving birth.

The crossword clue pops out in the loo

We have been finding it increasingly difficult recently to find out what separates the human species from other 'lower' forms of life. Those killjoy scientists point out to us that elephants have bigger brains, cockroaches live in more varied

Here is a technique I use to remember the list without getting out of bed. It involves painting a mental picture which will be memorable enough to last through till tomorrow morning. Create an image for each number up to 10 – something that rhymes with them, for instance:

ONE rhymes with BUN

TWO rhymes with SHOE

THREE rhymes with TREE

FOUR rhymes with DOOR

FIVE rhymes with HIVE

SIX rhymes with STICKS

SEVEN rhymes with HEAVEN

EIGHT rhymes with PLATE

NINE rhymes with WINE

TEN rhymes with HEN

Each of the items on your list is mixed with its rhyme to form a picture. Tomorrow I must remember to

1   Go to the dentist

2   Mend the chair

3   Take the car to the garage

4   Do the washing

5   Go for a jog

6   Practise for karaoke night

7   Set the video to record the match

8   Pick the car up from the garage

9   Watch the video of the match

10   Do the washing

Now all I have to do is paint the pictures and tomorrow the list will pour out of me.

climates, plants can count, dogs can understand words, crows can make tools, monkeys can understand what each are thinking, termites have a theory of mind. And if you want to know who really rules the planet – it's bacteria. Where can we turn for comfort? How can we claim superiority over other animals? I'll tell you: none of them can do crosswords. The cryptic clue was invented for the subtle brain. It is the best demonstration of the deep oceanic currents which roll beneath the surface of human awareness. The clue nags at you all day, then pops up like toast when you least expect it. This is why humans are the greatest.

SUMMARY

- Taking in the world cannot be just a matter of assembling information; there's too much of it. A large part of the task is slicing information away. The eye does it very well, removing 99 per cent of the information it receives. Mental structures do the rest.

- Attention mechanisms are very powerful tools for preventing information overload. They can allow us to zoom in on one particular dimension of the environment and ignore the rest. They can even cause us to turn the outside world off completely while we sleep, while still monitoring the environment unconsciously, making us wake up to specific sounds or smells.

- Problems arise from deciding which dimensions to attend to, and which to ignore. We build up a list of expectations, and search for those, shunting the other stuff up a siding. Dangerous practice – we are building up problems for later on, as we will see.

2

Getting the Measure

You've looked, smelt, tasted, listened and felt, but your knowledge of the world is still a jumble of neuronal hums. Now you must begin to slap the data around a little, prod it, get a feel for it; **measure** it. Is it big or small, loud or pale green? You'll be surprised how wrong you can be. Time and space are more warped in your mind than even Einstein could have ever dreamt.

Time flies when you're having fun

. . . and during Double Chemistry it crawls. The subjective measurement of time is capable of vast vagueness.

We have several internal clocks, ranging from the one that times our life-span – the tick-and-tock of birth-then-death – to the clock that judges tiny fractions of a second, which allows us to return a 100 mph tennis serve straight down the line. The clock mechanism that we use day-to-day is a loop of neurons running from the prefrontal cortex to the basal ganglia, on to the substantia nigra and back to the prefrontal cortex (see p. XXII).

One tick of a neural clock.

A nerve impulse travels around the loop at about 1 metre (3 feet) per second. By the time it gets back home, one tenth of a second has elapsed. This is one 'tick' of the brain's clock. So it takes, for instance, 3.5 minutes, or 2,100 ticks, to soft-boil an egg.

An interesting thing about this loop tells us a lot about time judgement – it works faster when it gets warm. In the 1930s a physiologist, Hudson Hoagland, noticed that his wife, who was suffering from a fever, kept overestimating time[1]; she would say he had been gone for an hour when he had only been away for 40 minutes. Being a scientist, he devised a number of experiments to do on her while she was ill. Being a very long-suffering wife, she agreed, and sure enough her estimates of time were about 20 per cent out. Unfortunately she recovered from her fever. Hoagland choked back his disappointment and cast around for more subjects. He wrapped heated coils around their heads and found that hotheads of any sort run their clocks about 20 per cent faster than others. They think time is running quicker than it is. Under the influence of adrenaline the same thing happens. Could this explain another common experience; people who are very anxious are always so afraid everything is going to be late?

How long a minute is depends on which side of the bathroom door you are on

(Zall)

As with waiting for a lift (see p. 189) there is nothing to do while staring at the bathroom door but cuss the door, cuss the person on the other side, and cuss Murphy, who kept the bathroom empty until 15 seconds before you needed it. Because there is a high urgency behind your need to get in there, your seconds will be a 'wee' bit faster than anyone else's. Hence the frustration.

Time, then really does rush by inside your head when you are in a heightened state of alertness. When at last the day has come to an end you remember the rush of time flying by.

In emergencies the neuronal clock can speed up a lot more. People often report that when they were in a crash everything seemed to happen in slow motion. In fact it is they who were thinking in fast motion.

In rare cases when the internal clock is disrupted by disease or accident, time takes a strange form. If the ticks are slowed due to damage in the system, a person will find the world rushing past at breakneck speed. Oliver Sacks, in *Awakenings*[2], describes a group of patients for whom time stopped. Encephalitis lethargica, a brain infection, had left them frozen, immobile. Treating them with a new drug, L-dopa, brought them back from a state which they described as being in a world where time stood still.

In youth the days are short and the years are long; in old age the years are short and the days long

(Nikita Ivanovich Panin, 1718–1783)

Old people find the years flit by in no time, and young things seem to whisk by at a frightening speed. Their internal clocks have slowed down. Aged relatives really are 'relatively' slow.

A watched pot never boils

(This saying was first recorded in 1848,
in Elizabeth Gaskell's novel *Mary Barton*.)

Time is certainly not rushing anywhere now there's a pot to watch. Quite the reverse. A day spent watching boiling pots, or staring at lift doors (p. 189) or standing outside the bathroom (p. 26), or waiting in a few queues (p. 176) can appear to last for ever. Here the anxiety which speeds up your clock makes three minutes seem like five.

Recollection of events like these seem to reverse the picture. The day that seemed too short when you were living it is remembered as a long one, packed with events, while the pot-watching, boring day leaves barely a trace, because there's nothing there to remember.

Short cuts are the longest routes

The words that send a chill down the spine; 'I know a short cut'. All passengers mentally add two hours to the journey time.

Any new route is a problem. You don't want to make a wrong turning and end up in Madagascar. Checking the route and remembering the landmarks every inch of the journey is brain-consuming. When you recall the journey later your memory is full of incident, and spiced with urgency and fear of getting lost. So it seems it was a long trip.

Next time you take the short cut it'll be easier. Anxiety will be less, the internal clock will be going slower. There will be less sense of rush. Incidents and landmarks will be 'chunked' – so that 'church hall, then petrol station, two sets of lights then poodle parlour and take the next left' will become 'left at the end of the High Street', a smaller, neater memory.

But it takes a few journeys for a short cut to start actually feeling like a short cut.

People like to plan short cuts into their routes. This is a hilarious idea in cities like London, where the no-entry signs and one-way systems have been designed specifically to thwart short-cutters. The roads are marked on the map, but there's no telling if you will be allowed to use them, so the London street map is about as useful as the I-Ching for getting you across town. The street signs will nudge you where they will. All the map can do is show you precisely how far off course you are.

This also explains another common Murphy experience . . .

Going there takes longer than getting back

The first time you take a route everything is new and different; everything goes into the memory banks. The whole cortex is buzzing with new impressions. The route back will be less fraught and easier to achieve, and will seem shorter.

You wake up two minutes before the alarm goes off

A fine example of the perversity of Murphy's Law. Just before the alarm is due to go off your eyes click open. You say 'brilliant, well done, I'm clearly alert and ready to leap out of bed, and I have two minutes to spare for a quick snooze'. You switch the alarm off and snuggle down for a moment. Next thing you hear is the fire brigade hammering at the door because they think you've died. It's tea time.

We have a natural circadian rhythm which wakes us during the day and puts us to sleep at night. It is different from the neuronal clock, above – it's a chemical one, controlled from the depths of the brain by the hypothalamus. The hypothalamus causes the pituitary to secrete the hormone melatonin at night. Melatonin causes drowsiness and sleep. The problem for the hypothalamus is that it is in the middle of the brain, where it is always dark. How can it tell if it has got the timing right? Is it day or night outside? It needs a window on the outside world, and it has managed to get one, by hook or by crook. It has hot-wired itself to the optic nerve, which runs just below it in the brain, and this way it can regularly hack into the activities of some interesting molecules in the retina.

There in the retina cells a pair of genes with the slightly suburban names of *TIM*

How the hypothalamus knows the time.

and *PER* are busily at work producing protein molecules called, surprisingly, TIM and PER. The production starts at noon, and the work goes on until the following dawn, by which time there are so many TIM and PER proteins that the *TIM* and *PER* genes are utterly smothered with them, give up and stop work. The TIM and PER proteins now start to fall apart, until at around noon the *TIM* and *PER* genes look around, spot that they've got space once more and begin making proteins all over again. The cycle has taken roughly 24 hours. ('Circadian' comes from the Latin *circa*, 'around' and *dies*, 'day'.) The interesting thing is that this reaction is light-sensitive, so it will time itself by the day's rhythm. As the hypothalamus eavesdrops on this saga, it can adjust its own cycle accordingly.

Even without outside light, circadian rhythm continues pretty automatically. In 1972 Michel Siffre spent 7 months underground, with no clues about the time of day outside. Even at the end of the stay, his sleep/waking rhythm was close to 24 hours. *TIM* and *PER* are necessary to avoid long-term build up of errors.

The whole thing goes widdershins when we travel abroad. Humans didn't evolve to go whipping around the planet, so the result is jet lag, when we drop fast asleep after lunch then bounce out of bed ready for breakfast on the stroke of midnight. Provided we sleep with the curtains open our *TIM*s and *PER*s can sort out the problem and get our circadian rhythm back to scratch, just in time for the flight back home, and more jet lag.

A man with one watch knows what time it is; a man with two watches is never quite sure

(Lee Segall)

Palaeolithic people had only one clock; the big round one in the sky. Their day was ruled by the sun, and there was no cheating; no switching on a false sun for a few extra hours in the evening. When the sun slept, so did they.

Early attempts at time keeping were not ambitious – early scientists tried to judge how long a year was. The Egyptians guessed 365 days. By the time of Julius Caesar they had upped that to 365.25 days, and the Julian calendar was set as world standard in 46 BC. Unfortunately it was wrong. A year is actually 365.242199 days long. So the Julian calendar was out by about 11 minutes every year. This is not bad, but not good enough. Gradually the official calendar began to creep ahead of the real one. Some 1,600 years later, in the sixteenth century, the folks in the field were seriously worried that the seasons were out of kilter. Which they were, by 11 days. In 1582 Pope Gregory had to admit the error, and 11 days were removed from the calendar. The folk in the field were furious, of course, with 11 days being stolen from their lives. They wanted them back right now (whatever 'now' might mean). Eventually they grudgingly accepted the new

Gregorian calendar, but Britain and America didn't catch up with the adjustment until nearly two centuries later. (This is why, although George Washington was born on 11 February, his birthday is celebrated on 22 February: the calendar was changed during his lifetime.)

Even after clocks were invented in the fourteenth century, time was a pretty vague concept. Nobody bothered to put minute hands on the clocks, for instance; there was no point, and no demand. Precision in timekeeping happened at a stroke when Huygens invented the pendulum clock in 1656. At last a device that could be accurate to within 10 seconds a day. Still the majority of people timed their lives by the sun and guesswork. This meant that noon was whenever you wanted it to be. Church clocks showed local time only.

The need for accurate timekeeping nationwide arrived in the nineteenth century, with the train. When people started turning up in Manchester an hour before leaving Liverpool, it was time to coordinate the local church clocks with the train timetable.

When it comes to measuring things, the football match is a special case. All normal frames of reference are cancelled, the real world recedes and a new reality dawns.

The last minute of the match lasts an hour

You are one goal up, there's one minute to go, they're in your goal mouth and the ball is going ponk, ponk, ponk against the woodwork. If only that devil-referee, who is clearly in the pocket of the opposition manager, would lift his whistle to his lips. Or even glance at his watch. How can a minute take so long? Well, now you know; it's because you're in a panic. Your internal clock is racing – his stop-watch has slowed to a crawl.

The other team's goal is smaller; the other team's goalkeeper is bigger; come to think of it, the other team is bigger all round

I've noticed something after years of being in teams, then watching my kids in teams; the opposition were bigger every single time. My small knowledge of maths tells me this cannot be. Not every single time. It has to be an illusion brought on by fear of the enemy.

To an active imagination the entire football game is full of inconsistencies. Apart from the obvious fact that their goalie keeps moving the goalposts, and the ref is blind, the pitch clearly tilts in their favour in the first half and the wind is in their favour in the second half.

Psychologists sometimes consider whether our picture of the world around is less like a photograph, more like a mediaeval icon, where important people are painted larger than ordinary folk, where nice people glow and float in the sky and nasty people are horribly hot and live in holes in the ground. Down here on the football pitch the enemy team are huge, heartless and threatening, and our plucky lads have mothers who love them. If the other side win it'll be the end of civilization, if we win, it'll be the triumph of truth.

If this picture is true, it would add an extra twist to size constancy (below), the ability of the brain to 'expand' distant objects to match up with nearby ones.

Do distant objects grow to reflect their importance? In a famous study[3] two groups of children were shown some coins and asked to guess their size. One of the groups of children were from a wealthy background while the other group were from a poor background. All of the children overestimated the size of the coins, but the poorer children overestimated more. This fits neatly with the idea that our minds reckon that size is secondary to significance, and adjust the picture accordingly.

The moon shrinks in photos

One of my more unpleasant duties is to remind you of the time you took a photo of the moon because it was so huge and beautiful in the evening sky, then when you looked at it you saw total black, with a tiny white dot in the middle. Why did the moon look so vast in the sky, yet so trifling in the picture?

This is the most absurd example of size constancy, a skill which you normally use to measure things at different distances from you.

When judging sizes the mind takes into account your distance from the object. If the image on the retina is small you might be looking at something huge on the horizon, or small in front of your nose.

Size constancy expands the mental picture of distant things. Normally you don't notice this, although that moon caught you out. Here's a simple example of size constancy in action below. The first picture shows two people walking along a pier. The second picture is the same, except the two people have been swapped.

This demonstrates how much your brain calculates the distances and adjusts the sizes to fit, using the perspective as a clue. The effect works even if the picture is replaced by abstract lines. In the third picture opposite the oblongs are the same size, yet your brain calculates that the right hand one is farther away and therefore must be larger.

The moon illusion works the same way. When the moon is high in the sky your brain automatically calculates its distance as about 200 metres (220 yards) away. (This seems to be the 'default' distance you ascribe to things of uncertain location.) When it's down near the horizon you realize it must be a lot farther away than 200 metres. Size constancy calculates it must be therefore much bigger than you thought, and that's what you see.

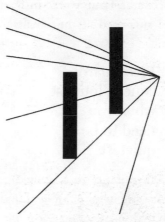

Ponzo's Illusion.

You always cook too much rice

As you gaze into the saucepan of rice your mind hits the skid patch. What will that lot look like when it swells?

Measuring volume is hell. It turns out that however clever our ability to measure distance and length, stepping into the third dimension is like leaping into the void. Try this: look carefully at somebody. Can you estimate how many times the circumference of their head will fit into the length of their body? Now measure it; wrap a piece of paper or a tea towel round their head, stretch it out and find out how right you were.

Or dig out a variety of mugs and glasses, then group together any which you think might hold the same amount of water. Pour water in them and see how you do. You'll be surprised at the results.

Floorboards squeak loudest when you need to be quietest

Let's not go into why you are creeping around the house when you think everyone else is asleep. Let's just admit that everyone has done it at some time, and been amazed at how exuberant floorboards become at dead of night.

The explanation points to how much you adjust to ambient noise levels. Daytime is very noisy, although you seldom notice this. But many of the sounds that fill your head during the day disappear at night – the traffic, street noises, air-conditioning, central heating, computer hum, dogs, helicopter gunships – all asleep now. It's quite surprising how much noise they make all together, and how much you acclimatize to them during the daylight hours. A creaking floorboard would be lost in the din at noon, but at midnight, in the silence, it sounds like gunfire.

The music is too soft for you and too loud for them

Nobody can scrape though adolescence without hearing those deathless words wafting up the stairs; 'WILL YOU TURN THAT NOISE DOOOOWN!!!!!' Once again the coffin-dodgers are moaning about your music. What can be done?

There are a couple of problems here. One is that you have habituated to the sound from your speaker. Your mind considers the volume to be normal, so that any quieter sound appears 'too quiet'. This habituation accounts for many of the strangenesses of our lives, not just aural ones but olfactory too. If you spend a long time in a room with others you get used to the smell you all make, so that you can't see what the fuss is all about when someone new enters and rushes to open a window. When 1970s astronauts came back from their record-breaking endurance flights in space, the ground crew must have had to draw straws to decide who was going to open the door of the space capsule they'd been sweating and farting in for weeks.

The second problem is that what might be intruding on their sensitive ears is not your volume but your taste in music. A sort of aesthetic habituation allows you to believe that drum'n'bass is beautiful, and allows them to believe that Mozart is gorgeous. Both sides of this partic-ular taste chasm will claim that the other's music is boring, repetitive and predictable, whereas theirs is the music of the spheres. Kin Hubbard said 'Classical music is the kind we think will turn into a tune', and you can see his point of view, although classicists would say that pop music is

the kind of music that never escapes from being a tune.

A handy solution to the problem is to turn down the bass on the speakers. It is the low frequency notes that travel through the house, while high frequency notes are absorbed by the walls of your room. Turn the bass down and the sound has a better chance of being confined to your room.

No matter how hot the shower, you are always cold afterwards

That uncomfortable shivering as you get out of a nice, long, hot shower can easily be remedied; don't have a hot shower, have a cold one, then it'll feel warm when you get out.

As the hot water cascades over you the heat sensors in

the skin tell the hypothalamus that you are too hot. The hypothalamus sends out a number of commands. One is to turn you lobster-pink – at least that is the side effect of increasing blood flow to the skin to lose heat at the surface. The hypothalamus also starts you sweating. Sweating is a brilliant use of physics; it takes a lot of heat ('latent heat') to make water evaporate. If that water happens to be sitting on your skin in the baking sunshine then the sun's heat is used to evaporate the water, not to bake you.

But the system evolved to deal with a Stone Age environment – hot, dry sunshine, not hot wet showers. In the shower, all the sweat does is add to the hot water. It is a superheated you which steps out of the shower. Any room, however warm, is going to feel cold under those circumstances. Soon afterwards you will be shivering, because at last the sweat evaporation will begin to take effect and cool you down.

SUMMARY

- The outside world has barely wiped its feet on the doormat of your mind and already you aren't sure who or what you're dealing with. For Palaeolithic people precision was not a top priority. They didn't measure things in millimetres, as we regularly do. Their counting system was simple, probably astonishingly simple; there are tribes today who have no words to describe quantities above two (see p. 182). Palaeolithic people couldn't be late for anything, because the planet is itself a bad time-keeper and produces Spring fiendishly early some years, then removes it a couple of months later with a cold snap, then gives half of it back.

- Nowadays we have to invent new words to keep up with our neurotically hyper-precise world. The freshly minted 'attosecond' is a billionth of a billionth of a second. Only the most fastidious of chefs will be using attoseconds to boil an egg. (If an attosecond was expanded to be a second long, then one of our seconds would last 30 billion years – twice the age of the Universe.) But if you want to time how long it takes an electron to move from one orbit around its atom's nucleus to another one, you reach for the atto-mic stop-watch.

- The optical illusions opposite are easy to print in a book, but the big wide world of illusions covers all the dimensions – time, sound, distance, size, volume, temperature, height; they're all slippery customers. And, as you will see, the next processes they go through don't always make things any clearer.

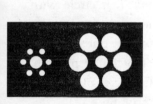

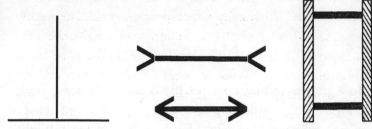

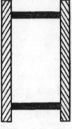

Which line is longer – the horizontal or the vertical one?

Which line is longer?

The two horizontal bars are the same size.

The central circles in each pattern are the same size.

The two squares are identical.

The white square is the same size as the black one.

3

Memories

How are we getting on with the jigsaw? When we look at a piece of wooden jigsaw we try to work out what's on it. A splash of blue might be sky, a hint of green might be grass, a circle with a dot in it might be an eye. To check, we look at the guide picture on the box. If the picture shows a rustic scene with people in it, then we were right about the sky, grass and eye. If the guide shows a computer room it's time for a rethink; that circle with a dot could be a button on a machine, not an eye.

In our mental jigsaw the 'guide picture' is our memory. But we've got a lot of memories; a lot of guide pictures. The circle with a dot might well be the barrel of a gun, a button on a shirt, a wedding ring, a symbol on a map . . . Which is right? That depends on our expectations. Here we enter a slightly murky world; we see, to a degree, what we want to see. Plenty of scope for Murphy's Laws here.

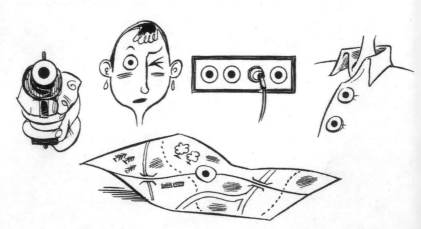

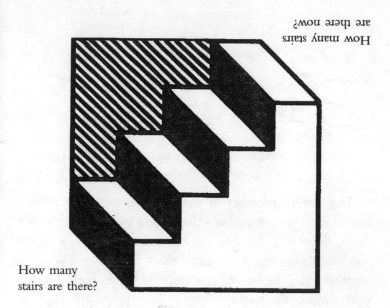

How many stairs are there? (upside down)

How many
stairs are there?

A simple trick shows this up. This piece of jigsaw shows a picture of . . . what? A collection of lines, that's all.

We see a flight of stairs because we have reached into our memories for what the picture reminds us of. Here is the optical illusion; turn the picture over and the stripy wallpaper moves from the back to the front of the stairs, and one stair goes missing. This magic happens because our only memory of stairs is looking downwards from above. We don't have any memory of stairs viewed from below, because we've never seen such a thing. So instead of saying 'here's some stairs . . . and here's the same stairs upside down', we say, 'here's some stairs . . . here's some different stairs'. This shows clearly that what we see depends on our expectations, and our expectations depend on our memories.

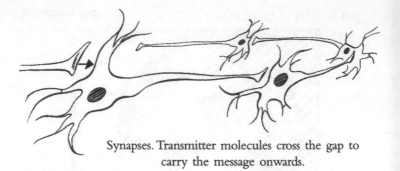

Synapses. Transmitter molecules cross the gap to
carry the message onwards.

The key to memory is the synapse, the point where
nerves join. 'Join' is perhaps the wrong word, because they
don't; there is a tiny gap between the end of one nerve
and the beginning of the next; that's the synapse. When a
signal reaches the end of one nerve it causes special transmitter molecules to jump across the gap, and the effect is
that the next nerve sends the signal onwards. Because the
nerves aren't actually joined, the connection is flexible.
This is the key to the rich complexity of human experience. If the nerves were joined like wires in a plug,
twisted or soldered together, there would be no possible
change from birth to death. Each human would behave
the same way all their lives. Although you may have the
impression that this is exactly how some people function,
it is not so. That slight separation means that the incoming nerve can change position, it can move closer to other
incoming nerves, or farther away. It can even drop off. On
the other side of the gap, the onward nerve can be just
as whimsical. It may decide to fire with one incoming
message, it may take several charges before it fires[1].

But there is one thing which all nerves seem to share;
once they have been fired off at a particular synapse, they

will tend to do it more easily the next time. In other words it has a memory of the last time it fired, which will cause it to fire again. It's called **'long-term potentiation'**, and it's what memory really is. When two things happen at the same time they tend to get hitched in the mind; as they say in the laboratory, 'neurons that fire together, wire together'.

The first time an infant eats an ice cream, they form a link between that strange new shape in their hand and that strange new taste on their tongue. The neurons involved are long-term potentiated; this is now a memory. The next time the child sees that same strange shape, the link to 'strange taste' will fire off, and they will remember the flavour once more, vaguely at first, but with each ice cream the link becoming stronger. A few ice creams later, the memory is fixed for ever.

In the dark there's always one more stair than you thought

Memories are stored in various places around the brain. Memories of how to do things (e.g. how to walk downstairs) are stored in the putamen. Memories of events (e.g. interesting stairs you have walked down in the past) are laid down in the hippocampus. Memories of facts (e.g. what stairs look like) are spread around the cortex. Some 'emotional' memories (e.g. encountering the last stair that shouldn't have been there) are burned into the amygdala.

It would perhaps be handy if 'fact' memory included an inventory of how many stairs there were in all the stairways you have been down. But come off it, we can't store all the details of every single thing. If we could quadruple the size of our brains, then yes, OK – cram five more kilograms of grey matter into the skull, then nothing could surprise us, certainly not the extra stair. Without that cranial loft extension we simply do not have the capacity to know everything. (We have the capacity to believe we do know everything, but that's another story.) Until then, the extra stair will come as a surprise.

Between one room and the next the memory is wiped

You're working on some complicated arty-crafty, gummy, mucky project on the living room table. You need one little thing from another room. You go into the other room, stop suddenly and stand there thinking, 'why am I here?'. Why does that happen?

While you were at the table your brain was full of a rich weave of sensations; the smell of the glue, the sounds of rustling paper, the touch of the materials, the light from the lamps, wafts of emotion as you steer through a series of little triumphs and disasters with the scissors and sticky tape. All the different components are based in different parts of the head (sights in the visual cortex, sounds in the auditory cortex, etc.) but they are linked together by nerves, like a loosely woven neural tapestry strung around the cortex (see also p. xxii).

Memories are collections of connections.

Here, in this new room, with its new colours, smells and sounds, there is nothing to remind you of what exactly you were doing; no link with the old memory strands. Therefore, oblivion.

Sometimes it pays to say out loud the name of the object you want as you set off for it. It is well known that saying a phone number out loud improves the memory of it. And it's easy to see why: the memory of a number you have just heard will be small (a couple of strands of tapestry, linking the auditory cortex to the speech recognition area close by, called Wernicke's area). Saying the number out loud involves more areas of the brain; decision-making, speech zones, motor cortex, and even the auditory cortex again, as you hear yourself repeat the number. So the tapestry will be bigger, the memory longer.

Front doors shut just as you realize you left the keys inside. They shut faster if you left them next to the umbrella you now need because it's raining

It is a sort of Zen moment; as the door to the house closes the doors of perception open. I believe this is closely related to the wiping of the mind between rooms. When you are indoors, the neural tapestry of sights and smells is a 'home' one. Outside you enter a 'travelling' tapestry, and different memories and associations (like needing the front door key) are sparked off.

In some households there is, just before the start of a journey, a moment of reflection just on the inside of the front door; the mind can relax, broaden its search, and imagine itself already on the way. Then last-minute things can be scooped up in time to avoid inconveniences, such as trying to call the fire brigade 10 minutes later on the phone which you suddenly remember you left next to the umbrella, which is next to the front door keys, which are inside the house.

Once you take a wrong turning, you keep on taking it

What makes this particularly irksome is that each time you drive up to the junction you remember that you took the wrong turning last time, so you try particularly hard to get it right this time, but to no avail; wrong you go.

The reason for this is a strange quirk of memory. As you approach the turning you try to remember what happened last time. You can dig out no memory of what happens when you turn left. However, the right turn looks familiar. You remember something interesting happening when you went down there. So, weighing a doubt against a certainty, off to the right you pop.

That vision of the fateful junction sparked off a memory of something interesting to the right. This association of two memories is called a conditioned reflex. In Russia 100 years ago Ivan Pavlov did a series of famous experiments on dogs. He rang a bell every time he fed them. Eventually they associated the two events and when he rang the bell they salivated automatically. Many psychologists consider the conditioned reflex to be the basic currency of the brain.

(Conditioned reflexes happen in the strangest places. Some friends told me they can't listen to a smoke alarm going off without feeling hungry. This is a peculiar response, but there is a reason: when they were young the sound of the smoke alarm was the sign that their father had nearly finished cooking supper. That slightly kinked conditioned reflex has remained with them ever since.)

For you, the conditioned reflex ensures that the more you turn the wrong way the more likely it is that you will turn the wrong way next time.

What you could really do with at that junction is a bit of aversion therapy. If at the very moment you turned right you got a powerful electric shock you would certainly remember not to do that again. On the next journey, as you approached the junction the memory loop would connect 'turning right' to 'hurts' – another kind of conditioned response.

Now, you might argue that you do get an unpleasant shock – the kids saying 'Dad, you idiot, we've gone the wrong way again' – but it happens later, as the road turns into a farm track. To be effective the shock has to happen at the same time as the sight of the junction, otherwise the link is not made.

Pelmanism: the same card shows up all the time

In Pelmanism, or Pairs, you lay a pack of cards face down then take it in turns to turn over two cards in the hope that they will match. Even hardened addicts find their hand straying to the same card again and again. The card becomes a kind of family friend, even though it loses them the game.

I believe (and I hope to be offered the research grant to find this out) that a conditioned reflex is set up around the bogus card, so when you look at it you retrieve a memory of it being in some way friendly or lucky, and that's what guides your hand.

Once you hear a new word you keep on hearing it

This is down to long-term potentiation – the tendency for a neural tapestry to be strengthened by repeated encounters. On page 4 we saw how few words are actually remembered from the hubbub around us. When a new word – like 'potentiation', perhaps – is brought to our attention it weaves a new little tapestry in the word recognition region of the brain, using long-term potentiation.

When the word crops up again the memory is rekindled. Suddenly 'potentiation' pops out of the babble and registers in your consciousness. 'Good grief, I hear that word all the time these days', you say.

(This is also known as the Blue Volkswagen Effect, after the man who tried to buy a car that would be easy to pick out in a car park. He thought a blue Volkswagen would be just the thing. Because nobody drives a blue Volkswagen, do they? As soon as he took it out on the road he noticed blue Volkswagens everywhere.)

The only tunes that you can't get out of your head are the ones you shouldn't have let in

After so many years of popular music, and only 72 notes to pick from, you would have thought all the tunes had been written by now. But nothing seems to halt the onward path of the irritating jingle. They pour forth from recording sessions by the bucketful. Thankfully only a few make it out of the front door, but these have been selected for mental tenacity, often using a technique called the 'Old Grey Whistle Test'. The doorman or caretaker is played the tune, then asked his opinion. The asking of the opinion is a mere formality; what really counts happens later. The old grey-haired fellow is secretly watched as he leaves the building. If he is whistling the tune, then it must have lodged in his head. This is the first and major test for a hit song.

When you search for something, you go back to the same place again and again

There comes a time in everyone's life when we say 'I hope nobody is watching me; I feel such a fool'. We have just looked for the tenth time in the place where we left our book, and we know it's not there, but now we're going back to check once more. And no, it's still not there, so we'll just look once more to be doubly certain. Why do we keep searching the same spot when we know it's not there? Because on the twentieth look, there it is!

The mystery of the invisible object has puzzled people for centuries. In recent times quantum scientists have wondered if the object is slipping into a parallel universe, where it stays until we buy a replacement. Two thousand years earlier Archimedes had an answer based on the curious science of his time. The theory was that vision was an active rather than a passive process. Instead of letting light beams into the eye, vision was conducted by sending beams out, like searchlights, to scour the area. If the search beams were angled in the wrong way they wouldn't pick up the object.

The truth is now known. When searching for the book we don't stare hard at everything, we take a short cut; we root around the memories in the cortex for an image of what we are trying to find. Keeping that in mind we rapidly scan the room looking for something that matches it. It's a clever trick.

The trouble happens if we are in a hurry. In the hope of a quick win we scan for a narrow range of options – perhaps the distinctive red of the book cover – and simply don't notice anything else. Which is a problem, because if we left the book face-down, with the cover hidden, the match will never happen.

The solution is easy; go and make a cup of tea. While we are distracted by tea bags and milk, the mind broadens its search. When we go back now, instead of just the flash of red, we look for any kind of 'book' image. The match now happens rapidly, and we're left wondering why we never spotted it in the first place.

The noodle effect: when at last you have to ask the supermarket assistant where the noodles are, they're right beside his head

A surprisingly frequent occurrence. The thing appears at the most embarrassing moment, so it seems. You searched for ages with your noodle image in mind. When the time came to give up, your mind relaxed, the notion of 'noodle' broadened, the packet popped into view. Suddenly you were the noodle.

The Noodle Effect.

This happens so often to so many of us that you might be tempted to think the nation's brains have started to degenerate. But our brains evolved to live among trees, rocks and goats, not specially reduced three-for-one bargain bins. Palaeolithic plants don't compete for our attention with the same zeal as shampoo manufacturers, rocks don't come in the same variety of competing designs as breakfast cereals.

A heavy, ugly, useless thing has sat in the dark corner of a cabinet for 20 years. The day after you throw it away, you need it urgently

Ah, the wonder of the glory hole, full of things that might come in useful one day. All forgotten now, because the film canister, jar of curtain hooks, packet of fuses and cable ties were plonked in there quickly, without a memory trace getting a chance to form in your mind. Consequently everything was forgotten. Only the act of throwing it away causes a memory to form. Too late, of course.

Déjà vu has to be seen to be believed

That extraordinary feeling you get occasionally (nearly everyone has had it) that you've been here before seems to suggest there is another you in a parallel universe, or maybe you were here in a previous life, or maybe you have powers to foresee the future. Or maybe, just maybe, there's a problem with your time tags.

One of the most important aspects of your dealings with the world concerns time. Everything that happens to you (such as you reading this book) is first laid down in the hippocampus as a story set in time, 'right now . . . a little while ago . . . earlier than that . . .' You remember, for instance, that you read these words more recently than the ones above. The hippocampus gives everything a time tag.

The time tagging is essential. If you had no idea what order things were happening, it would be like watching a film in which the frames have been assembled at random. It would completely ruin your day. In fact the idea of 'day' would be meaningless. The sun would be dodging all over the place in your memory. Your experience of each moment of your life is therefore fixed in a mental time chart.

Simultaneously you are comparing each new experience to memories and expectations already stored in the brain. If your experience of a particular moment, as it travels forward to be edited into the film, fails to pick up a time tag, the cortex assumes that since it doesn't fit into the present, it must be from the memory section – it must belong somewhere in the past. Where in the past it can't say, so déjà vu experiences have fuzzy histories.

MEMORIES AND FORGETORIES

How happy we are when we can just forget about our troubles. It is as important to forget as to remember. Schopenhauer said, 'to expect a man to remember everything he has ever read is like expecting him to carry about in his body everything he has ever eaten'. Even enormous pain, trauma and helplessness has to be allowed to fade in certain cases, otherwise mothers would refuse to have more than one baby.

Memories have different time spans. Some are indeed short: working memory, which deals with the job of the moment (in my case, the typing of this paragraph), uses short-term memory to pick up information, then drop it into oblivion a few seconds later. At the other extreme, some long-term memories last all our lives. Things which have been practised weave a stout tapestry – we never forget how to ride a bike, or the two-times table.

We forget our pleasures, we remember our sufferings

(Marcus Tullius Cicero, 106–43 BC; Pro Murena)

The problem is what to forget, and what to remember. It is really rather useful that we remember some of our sufferings. If we remember bad things of the past we have a chance of avoiding them in the future. Many traumas remain 'etched in the mind' forever. (For instance, 'flash-bulb' memories – we all remember where we were and what we were doing when we heard about 9/11, the attack on New York's World Trade Center.) These have been stored in a special location – the amygdala, a tiny lump in the heart of the brain. The amygdala is the seat of our most basic emotions, particularly fear. The brain allows ordinary memories to perch around the cortex, exposed to the eroding currents of everyday traffic around the mind. In time they wear away, but memories of especially life-threatening events must be retained, therefore the amygdala stores them close at home for instant access. While ordinary memories fade like ink stains on the skin, traumatic memories are like tattoos; permanent. These are the ones which Cicero refers to above.

There is some argument about how much memories do fade. Oliver Sacks, in *The Man who Mistook his Wife for a Hat*, tells of an elderly woman who, under the influence of the drug L-dopa, recalled songs, jokes and memories from 40 years previously, memories she thought were gone for ever. Certainly there are billions of memories up there in our loft, though many of them are pretty worn by now. If they are to be recalled they will need restoring with fresh threads. See p. 76 for what happens then.

It is also a good idea that memories can fade. The Russian scientist A.R. Luria, in *The Mind of a Mnemonist* (1987), described one individual, Solomon Veniaminovich Shereshevsky, who could remember vast lists of random words, and retain the memory over several years. He even made a career out of it for a while. But before we feel too strong a pang of jealousy about this, those of us who can forget a telephone number between putting down the phone and picking up the pen, we should realize that Solomon had a mind constantly bubbling up with images of things, words, tastes and colours, which continually intruded on his normal life. He could have done with a better forgetory.

You remember the face, but the name evades you

Names are never as good as faces at sticking around in the brain. A large part of the visual cortex is devoted to face recognition. In fact the overspill from this hyperactive face-recognition department causes us to see faces in clouds, tree bark, among the rocks of Mars, etc. Names, however, are not so lucky. Language is a modern development, less than 100,000 years old; the brain hasn't built as big a department, so our common experience is that we 'always remember a face, but the name evades us'.

Absence makes the heart grow fonder

(Thomas Haynes Bayly, 1797–1839, song writer and playwright)

We aren't built with perfect memories. As time goes by the tapestry fades, leaving behind a few threads, a summary which, when recalled later, gets all the detail stitched back in.

Our wooden jigsaw might fade over the years – the colours wash out. All we have to do is paint back the colours to restore the picture. But can we be sure we are using the right tints?

Likewise, as we restore our mental tapestry, are we stitching in the right threads? As we go to greet our long-lost friend our memory is fond, and that influences the threads we use. So we'll be greeting the warm and happy memory. It'll be some days later that we will be reminded how we used to suffer their snoring.

Absence diminishes small loves and increases great ones, as the wind blows out the candle and fans the bonfire

(François, Duc de La Rochefoucauld, 1613–1680)

When we restore the tapestry we use only loving threads for the loved one, and exclusively nasty threads for the one we were not quite so fond of. The picture becomes skewed, caricatured.

On p. 76–7 we will see that the brain can recall memories incorrectly, or make up memories that were never there, how in some cases 'absence makes the heart grow wronger'.

But by and large, Albert Schweitzer was right when he said, 'Happiness is nothing more than good health and a bad memory'.

SELLING LITTLE WHITE ONES

Because of the splendidly exuberant way each of our 10,000,000,000 neurons work, with all their fuzzy connections, memories are bendy. Nobody understands this better than an advertising agency. With finely honed skills they gently warp reality enough to allow you to drop yourself in it time and again.

THIS BOOK WILL SAVE YOU MILLIONS! £99·99 "HOW TO SPOT A RIP-OFF!"

HOW TO SPOT A RIP-OFF

Franklin Pierce Adams affirmed in 1944 that 'There are too many people who believe, with a conviction based on experience, that you can fool all the people all the time'. This section proves him right.

Everything is a penny short of a sensible price

I was in a wine shop once. Without moving an inch I counted just over a thousand 9s. Every bottle had a big label on it: £2.99, £3.99, £4.99, £5.99 or £6.99. You see, three pounds is three pounds, but two pounds ninety-nine is two pounds, plus a bit. Our memory tells us that '2 . . .' is the same as '2', we don't think, 'just under 3'. How strange it is that we know the trick that's being played, but they continue to play it. And we, bless us all, continue to willingly fall for it.

A trade apologist once gave an explanation for the 99p phenomenon. 'It's an honesty check,' he said. 'If the price was £5 the sales staff could easily pocket the money without anyone knowing. But because they have to give 1p change they are obliged to use the till'. Nice try, Mr Apologist.

The more buttons and flashing lights, the less it can do

We find here a phenomenon known to psychologists as the 'supernormal stimulus'. When we see the flashing lights and sleek lines a memory is sparked off from science fiction films of space ships with superior intelligent, high-performance, multifunction, trillion-dollar quantum computers on board. A small part of our brain should be calculating that if so much budget has been spent on the facade, there can't be that much money left for making the interior function properly, but this notion is overwhelmed by the more alluring memory inspired by the twinkling lights.

Supernormal stimuli were investigated by Niko Tinbergen, one of the founding fathers of modern psychology, who devoted his life to the study of gulls. One of his observations was that if a false egg was added to a nest it would be treated as one of the family. Most strangely, if the egg was huge, mummy gull didn't guess it wasn't one of hers, she lavished more care on it. It was **super-egg**.

I find the same effect with my children watching TV soap operas. The characters are so much more competent, better dressed, prettier and wittier than real life. When a child turns from watching their favourite soap back to the real world they find their family unbelievably scruffy by comparison; they don't know their script, they are badly lit, not accompanied by the right music, and fail to have their hair and make-up touched up in between shots.

Packaging has the same super-normal tendency. If it looks big, it is big, says our brain. A big box means a big object inside.

The supernormal stimulus is an example of evolution gone a little bit off the rails. Take also the peacock's tail.

All sorts of excuses have been offered for this extraordinary growth on the back end of a peacock. From a basic survival point of view it's a disastrous piece of baggage to have to drag around. But the poor fellow won't find a mate unless he has the avian equivalent of our flashing lights and bling-bling.

Certainly, manufacturers will exploit the supernormal stimulus effect to the hilt. As Stephen Leacock pointed out, advertising is 'the science of arresting human intelligence long enough to get money from it'. The still small voice that mutters 'you can't tell a book by its cover' is drowned out by the scream 'first impressions count! Buy it'.

The first law of advertising is to avoid the concrete promise . . . and cultivate the delightfully vague

(John C. Crosby, American Civil War diarist)

. . . Thus ensuring that the vague promise is enhanced by memory traces that promise more. When you buy a 'three season' tent you little realize that the seasons they refer to are early summer, midsummer and late summer. A 17-inch television screen is only 17 inches if you measure diagonally.

Some products proclaim a 'new, improved original formula' offering the best of both worlds, ancient and modern. 'Natural ingredients' often include 80 per cent natural sugar, fat and salt. 'Organic' sounds good, but for some producers it means simply 'grown'. 'Succulent' sparks off a delicious memory – what it means is 'packed with fat'.

The bigger the bargain, the more money leaves your pocket

Curious, how you can spend all day in the shops saving money, and still return home bankrupt. It's all 'Big Savings!', '0 per cent finance!' and 'Massive Reductions!!' Each encouraging sign sparks off the right kind of memory; '0 per cent finance!' seems to be telling you the item is free. I've never yet seen the sign that goes 'Save up to 100 per cent on this exclusive sofa-bed! Don't buy it!' Thanks to the advertisers, aided by our vague memories, we come away with anything we are asked to. The advertising copy definitely works; a simple experiment by Nisbett and Wilson[2] showed how much. They asked women to judge between different pairs of stockings. The stockings were all the same, but the subjects were told they were different. Asked to select which they thought was best, they nearly all managed to select one pair which was 'superior' for any number of reasons; colour, texture, quality – it was all make-believe, but it was real enough for them.

There's a sucker born every minute

> (Phineas T Barnum, of Barnum and Bailey's Circus,
> America's most famous showman)

During the mid-nineteenth century Barnum's extravaganzas toured the world, featuring a variety of unlikely but hugely popular freaks. Barnum's problem was what we would nowadays call 'throughput': the crowds loved it so much they hung about, making it impossible to cram more punters into the tent. Barnum had a sign placed at the far end of the tent which read 'Egress'. The word sparked off a vague memory in the punters' minds: wasn't that some sort of exotic bird? Eagerly they crowded through the door, and found themselves, of course, outside.

SUMMARY – THE ONCOMING TRAIN

You're running at breathtaking speed along a tunnel. It's very dark, except for a point of light up ahead of you. Is it the light at the end of the tunnel? In the Roadrunner cartoons exactly this question goes through Wile E Coyote's mind as he chases Roadrunner through the tunnel. His memory tells him that tunnels have light at the end. It ought to be the light at the end of the tunnel; for anyone else it would be . . . But another memory, from past tunnels, says it is never that easy. It might – just might – be the headlamp of an oncoming train . . . Sure enough, Wile E Coyote is obliterated by the train.

How often does that happen to us, too? We rely on our memory to tell us what we're looking at, but the memory might be wrong, and we end up obliterated by the oncoming train.

4

Making Connections

So far we've taken in our jigsaw pieces, we've measured them, we've reached into our memories to see what they remind us of, now we have to fit them together to form the picture. In this chapter we see how, in the Murphy world, we regularly make the wrong connections, pick up the wrong piece, which refuses to fit snugly into the gap that's prepared for it. But rather than find a piece that does fit, we doggedly cram in the piece we've got, squashing it until it stays, or morphing the surrounding pieces to allow it in.

NAIVE SCIENCE

All we want to do is make sense of everything that's happening. To write a believable story from what we see, we employ **naive science**. This means applying the laws of cause and effect; everything that happens is caused by something, and then causes something else to happen afterwards. The problem is working out what is the cause and what is the effect, and it's here that Murphy's Law thrives.

To help us work out the cause and effect of events we have an armful of expectations, based on basic beliefs about the world. In Stone Age days the expectations were pretty accurate, because the world was easier; what went up, came down. Nowadays that's no longer a certainty. We crave pure, simple truth, but as Lady Bracknell said,

The truth is rarely pure, and never simple

(Oscar Wilde, 1854–1900, *The Importance of Being Earnest*)

Here are some examples of naive science.

People who wear hats go bald

The hat caused the baldness, QED. In fact it's the other way round: people who go bald wear hats because the top of their head gets cold. But the naive scientist doesn't believe heads can get cold. (He has a fine head of hair himself, of course.) He also has a vague notion that things that are covered over tend to die. Plants and animals suffer, why shouldn't hair? Hats off to the naive scientist!

Cold weather causes colds

The story goes: Colds happen more in winter, therefore it's the cold weather that brings them on. Not so: Cold weather causes us to cluster together in warm places, and bacteria and viruses are more easily passed among us. (Although if you don't wrap up warm, your immune system is suppressed and infections can get a grip on you.)

The wind is caused by the waving of trees

Several bright children have made this connection. We can laugh at the childish simplicity, but only if we forget that an eye-blink ago we adults believed the sun went round the earth, that we could make anything by mixing the right quantities of earth, water, air and fire, and that humans were a completely separate entity from all other animals. For thousands of years, since Hippocrates and Aristotle, we absolutely knew that the heart's function was to attract blood rather than pump it, that the lungs were there to mix blood up, that the brain's function was to cool blood, that vision was caused by beams of light emanating from the eyes . . . The people who were in charge of our health were thinking completely back-to-front. With health care like this it's a miracle the human race survived the Middle Ages.

Even now most of us think heavy things fall faster than light things, that loud noises travel faster than quiet noises, that the moon is bigger when it's near the horizon.

Currently all top scientists believe that the Universe was created in a Big Bang. The cause and effect of this seems undeniable. In the 1920s Edwin Hubble noticed that all the stars in the Universe were receding from us and each other – the Universe was expanding. Running the clock backwards produced the only possible cause; long ago all the stars and galaxies must have been concentrated at one point, a singularity, from which they exploded outwards to form the present Universe. It's a beautiful story, and one which I believe utterly. Brush aside recent discoveries that Hubble's theory can only account for 5 per cent of the Universe, that the Universe is expanding faster than it ought to be, that 95 per cent of matter is unaccounted for, forget dark matter or dark energy. I believe in the Big Bang. I'm happy with my jigsaw as it is, thank you.

Car washes take your brakes off

Murphy says 'gotcha!' to that one. As the brushes of the

car wash start to spin around you, the car starts to slide forwards. You scrabble for the brakes, convinced that you're about to plunge through the front wall of the car wash and die. At last you spot that the car isn't slipping forward, it's the brushes which are moving backwards. The naive scientist in you believed that the huge brush machine was stable, therefore the cause of the movement must be the car.

It's the same, but different, on train journeys; when you have settled comfortably on the train, waiting for it to set off, and suddenly the station starts to go in reverse. In this case the naive scientist assumed that the train was stationary. If the train was going to move, said he, it would jolt. Without the jolt the only cause of what he could see out of the window was the station's movement. You can see how apt the label 'naive' is.

Hot under the Collar

Professors Elton Mayo, F.J. Roethlisberger and William J. Dickson have an unenviable place in the pantheon of psychological research. Theirs is the most famous experimental banana skin since the invention of psychological research[1]. They spent eight years at the Hawthorne Plant of the Western Electric Company in Cicero, Illinois between 1924 and 1932, studying how to get the most production from the workers. They tried raising the temperature of the workplace by 2 degrees – production went up. They raised it by a further 2 degrees – production went up more. Up 2 degrees more – up went production to match. They were ready to conclude that keeping factories at sub-tropical temperatures was the best way to maximize production.

To finish off, they dropped the temperature back to its original level – production went up once more! The same thing happened when they studied light levels, or group size, or shift length. With each modification production went up. Eventually, grudgingly, they came to the conclusion that the workers were increasing production not because of the changes to their working conditions, but because they believed that somebody cared. Somebody was going to the trouble of finding out what their favourite temperature, light level, group size and shift length was. They were responding to consideration and respect.

Messers Mayo, Roethlisberger and Dickson had been trained in laboratories where they were working with rats. When they looked at the Hawthorne workers all they could see was rats – better dressed, sure, and able to walk on their hind legs, but apart from that, rats. To find the workers thinking such sentimental, wishy-washy, un-rodent thoughts took them by surprise.

For Mayo, Roethlisberger and Dickson the expectation was that the jigsaw would fit together a certain way, and it was a long slow education that taught them otherwise. At least they had the decency to own up, for many others that seems to be impossible.

Enough research will tend to support your theory

(Arthur Bloch, *Murphy's Law*, 1977)

This is especially true of drugs companies who fund many trials to test the worth of each new diet aid or antidepressant they wish to launch. Many of the trials are not published by the company. Of those that are, an unusually large number are favourable. And as it turns out, yes, the unfavourable tests are being gently but firmly smothered[2]. The research company is involved in cause and effect. The 'cause' is the drug; the 'effect' should be a recovering patient. If that doesn't happen, then it's the patient that's wrong not the drug. Trials continue until a patient recovers after taking the drug, then the company rushes to the printers.

Seeing is believing

We fancy that what we see is based on rational observation, but because of our suitcase of expectations we are much more likely to see what we want to see.

In December 1901 Marconi was ready to test his theory that radio signals could be transmitted right across the Atlantic. He sat in a tiny room in Signal Hill, Newfoundland shortly after noon, waiting for the signal from Poldhu in Cornwall. As he reported it later, he heard the letter S in Morse 'tak . . . tak . . . tak' clearly in his headphones. Modern analysis of the weather conditions at the time cast doubt on the claim. Medium-wave signals of the sort he used could not have travelled that far on that day. It is most likely that he heard a mass of static, and imagined the signal. He heard what he wanted to hear.

As it happened Marconi was right about radio signals, but many claimants are not so lucky . . .

Believing is seeing

At about the same time that Marconi was 'hearing' radio transmissions, René Prosper Blondlot, a French physicist, was 'seeing' a new type of radiation. The 'N-ray' was created by refracting X-rays through a prism made of aluminium. It showed as a very faint glow, so faint that only those with very sensitive eyes could see it. Foreign scientists were sceptical, but many eminent French scientists visited Blondlot's laboratory and witnessed his N-rays, which is strange, because they didn't exist. This was the emperor's new clothes in modern drag. The eminent scientists saw what they wanted to see. The folly was revealed when a visiting journalist secretly removed the aluminium prism from the apparatus. The scientists continued to marvel at the glow right up to the moment when the prism was dangled under their noses.

A lie can be half way around the world before the truth has got its boots on

(Mark Twain, 1835–1910, American author)

Truth is complicated, but a lie is streamlined. All the credibility of a lie is there because the uncomfortable dangly bits are tucked into the weave of plausibility. Nice smooth lies are easily swallowed. A lie will also fit in with our expectations. We like to construct a picture of the world as simple as a pantomime plot, as the stories below illustrate.

• In 1929, when Wall Street crashed, many people committed suicide in despair . . . Not so, I'm afraid. Journalists heard of one or two suicides at about the same time as the crash and constructed a cause and effect around it. Hospital statistics show no measurable increase in suicides during the months surrounding the crash. Statisticians are real killjoys like that.

- In August 1966 a ten-day blackout hit New York. Nine months later there was a sharp increase in the birth rate. If you want to believe this story, don't look at the figures, which show no spike at all, in either sense of the word. The Blackout Baby Boom is a splendid urban myth.

And here's a rural myth . . .

- In 1958 Walt Disney dispatched a film crew to Canada to film lemmings throwing themselves off cliffs, the way they do. After a couple of days wait, during which the lemmings showed an unexpected tendency to keep away from cliff edges, the impatient crew corralled several dozen of them, pushed them over the edge and filmed the result. The film, *White Wilderness*, perpetuated the myth of the suicidal lemming.

  Walt Disney knew how the jigsaw was supposed to look, and made sure he didn't thwart his audience's expectations. If you watch *White Wilderness* today it is clear that the lemmings are being propelled from just off-camera towards the precipice, but the naive audience will only see what they want to see. The modern world is rich and strange, but our Neanderthal minds prefer the simple things of life, or in this case, death.

They always shoot the messenger

We are all 'naive scientists'. Even warlords, tyrants and dictators are, although they lean perhaps less on the 'scientist' and more on the 'naive' side. A naive scientist sees two things occurring at the same time and assumes that the one causes the other. The area of the brain responsible for this intelligence is the frontal lobe. We must wish for future evolutionary changes to favour that part of us in aeons to come, particularly in warlords, to spare them from some of their more preposterous ideas. Meanwhile, when the messenger brings bad news the warlord will think 'bad news . . . messenger . . . shoot the messenger'.

We all do a little naive science, however hard we try not to. We can't find that cream bun we bought yesterday and our four-year-old daughter has gone missing . . . We tend to link the two and guess that the one ate the other. We can't find the car keys and our nineteen-year-old son has gone missing . . . We can't find our wife and our best friend has gone missing . . . Even in the best-regulated homes there's room for a momentary doubt.

THE BACK-STORY

Life can only be understood backwards, and can only be lived forwards

(Søren Kierkegaard, 1813–1855, Danish philosopher)

It gets worse. You may have thought up to now that, whatever mess we make of the present, at least the past is safe from harm, locked away in secure memories around the cortex. No such luck.

Hindsight is 20–20 vision

(Billy Wilder, 1906–2002, American film director)

Both Billy Wilder and Kierkegaard were wrong, as it turns out. We rewrite the past to fit in with the present. Hindsight is 20–20 **re**-vision. Nader and LeDoux[3] found that well-established past memories can be abolished almost with a flick of the fingers. They were working with rats in laboratory conditions, but human experience is that we can not only abolish the past, but can substitute a completely new one if we feel the need.

Examining the human brain, we find this is not hard to do. Memories are strung out around the cortex in a number of sites: the sound parts are in the auditory cortex, sights in the visual cortex, matters of geography in the hippocampus and emotional flavouring in the amygdala. The different segments are linked by neuronal connections, like a loosely woven tapestry. Over time each thread is subjected to selective decay. Retrieving the memory involves teasing out what is left of the tapestry from its perches around the cortex, then embroidering the details back in.

While some parts of the tapestry don't fade – the fear response in particular – other parts can be almost completely lost. So the reconstruction can easily be a distortion. Eyewitness accounts of crimes are notoriously unreliable because of this.

A series of examples of falsely recovered memories erupted in America during the 1980s. Nurse's assistant Nadean Cool had psychotherapy in 1986 to deal with a family problem. Tutored by the psychiatrist, she recovered memories that had lain repressed for decades, which would account for her psychosis. She remembered having been in a satanic cult, eating babies, being raped, having sex with animals and witnessing the murder of a friend when she was eight. This alone would be enough to account for any syndrome, but the psychiatrist helped her recover more memories. She discovered to her surprise that she had more than 120 personalities – children, adults, angels and even a duck. At this point she began to doubt her own insanity.

Eventually the manipulative techniques of the psychiatrist were laid bare, but not before many other cases of false recovered memory had been discovered and many families devastated by similar 'revelations'. Clearly these memories were false, but how easily had the preposterous ideas become part of their official biography! The victims could not believe that the events they were recovering were untrue, because they believed in the sanctity of the past and the purity of their memories.

Nostalgia ain't what it used to be

We should suspect the rosy glow of the past; it is being seen through the jaundiced eye of the present. The past was a laugh. It was carefree, money was easy come, easy go. It was warm and sunny. There was no robbery. Nowadays things are going too fast, crime is on the increase. The youth of today have no respect, they don't listen . . .

Truth is, what has changed is not the youth but the critic. Older folk are frailer and slower, less capable of coping. They feel the cold more, so their youth seems warmer by comparison. Their internal clocks have slowed (see p. 27), so the youth of today seem to move at bewildering speed; they brush old folk aside. But the old folk should remember that when they were young they did the same to their old folk.

Love is a dirty trick, played on us to achieve the continuity of the species

(Somerset Maugham, 1874–1965, English writer;
A Writer's Notebook, 1949)

In matters of the heart, passion and pain are felt much more acutely. With the end of a love affair comes hyperactive rewriting of the back-story. When She finds He has been playing away from home, the history of the previous six months is revised: 'So that's why he was back late . . . so that's what the 'bruise' on his neck was . . . so that's why he was so tired in the evenings . . .' The revision is long and thorough. The angel is repainted as the devil. This may involve a bout of post-traumatic stress disorder – sleepless nights spent reliving the past and rewriting the back-story to fit the new facts.

NOT MAKING CONNECTIONS

We have seen how easy it is to make the wrong connections, to get the causes and effects in a mess. There are also dire consequences from *not* making the *right* connections, not seeing the effect of a cause.

Whatever you like is illegal, immoral or fattening

(Alexander Woollcott, 1887–1943, American writer and critic)

Overeating is a global problem which we don't seem to be able to master. The brain we have brought with us from the past hasn't changed much in 100,000 years. Our ancestors ate leaves, roots, nuts and berries, and hunted when they could. Often the hunters came home with nothing - fast food was fast in the wrong way back then, and not so easy to catch. It was a hard life, no doubt, and brains were built to deal with it; they made early humans eat when they could because often they couldn't.

In the modern West we eat, as we always have, because you never can tell when the next famine will happen. But our brilliant brains have arranged things so the famine will never happen. The unfortunate side effect is that a third of North Americans are officially obese, and the rest of the world is swelling to match. Still we nibble, peck and browse, because that's what we are built to do. Something should be making us stop eating, but it isn't happening. We wait for some instinct somewhere to kick in and tell us not to feel peckish any more.

There actually is something that is supposed to do that job. The hypothalamus, deep in the centre of the brain, monitors the blood for sugar levels. When the blood sugar is high, it registers satiation with the higher brain. But somehow the higher brain fails to do anything about it. The connection is not being made, because our brain is not appropriate for the world it has created. This is

Morphy's Law; we are morphing into caricature humans because at a deep, instinctual level we cannot make the connection between eating and blobbing out.

You can only predict things after they have happened

(Eugene Ionesco, 1912–1994, French playwright; *Le Rhinoceros*, 1959)

Governments and scientists have the task of making predictions and preparing policies to cope with it. Rapid changes and innovations in science show up cruelly how hard this is. When the magnificent new molecule, chlorofluorocarbon (CFC), was invented, industry celebrated. It was perfect; cheap to make, non-polluting and non-toxic, a gift to the aerosol and refrigeration industries.

Nobody foresaw what CFCs would do to the planet. How could they? The molecules had to rise into the stratosphere and undergo dozens of subtle reactions up there under the influence of ultraviolet light before they began to destroy the ozone layer at reckless speed. No human could have seen that far into the future. Unfortunately it took a long time for anyone to see clearly into the past either. The comfortable story of the miracle molecule was knocked by the discovery in 1984 of a hole in the ozone layer caused by the profligate habits of CFCs. Even though scientists were all agreed on the severity of the damage and the threat to life on the planet, it took three years for an international ban to be instigated, and even now the ban is not completely in place. China, for instance, will not have phased production out until 2010.

When you make your face cooler you make the planet hotter

Climate warming, like obesity, is something we can see coming racing for us, but we can't make a connection. During the long hot summer we put on a fan to keep cool. The fan gets its energy from a power station, which pumps carbon dioxide into the atmosphere as a by-product of electricity production. Carbon dioxide is a greenhouse gas; it makes the climate warmer. When we feel warmer we put on a second fan to keep cool. Two fans use twice as much electricity, which produces twice as much carbon dioxide and therefore twice the global warming. So we're even warmer, so we put on another two fans . . . So it goes, and so go we.

The wild gyrations of the climate, from severe drought to hurricanes and floods, are caused by the extra heat in the atmosphere. The evidence is clear, the science proven, but our Palaeolithic love of cooling breezes continues. In the end global warming will fry us up, unless we make a connection between air-cooling devices and the air-warming effect they have.

SUPERSTITIONS

The inverse of cause and effect is cause and **no** effect. If you do something and as a result a disaster is avoided, you will naturally continue to do it. This is the nature of superstitious acts.

We avoid the number 13, just in case it is as unlucky as others say. Nearly all office blocks and hotels lack a 13th floor or a room 13. Many airports skip the 13th gate.

Aeroplanes have no 13th row of seats. Italians have no number 13 in their national lottery. In the streets of Florence, the house between number 12 and 14 is addressed as 12½. Many cities do not have a 13th Street or a 13th Avenue.

In France, socialites known as the quatorziens (fourteeners) once made themselves available as 14th guests to keep a dinner party safe from fate. Nobody will allow the number 13 into their life, even as an experiment to find if 13 is unlucky.

The tenacity of superstitions lies in the 'proof' that by performing the act, disaster is avoided. When you spill salt you chuck a pinch of it over your shoulder just to be on the safe side. And the result? No disaster! QED. You will keep on doing that to be on the safe side. You wouldn't want to tempt fate, would you?

If you're scoffing at the salt superstition, ask yourself if you use only lower-case letters in e-mail addresses, just in case. E-mail addresses are not case-sensitive, but would you ever try sending one with capitals in, just on my say-so?

SUMMARY – THE MISSING LINK

Whether we are top scientists, amateur scientists or non-scientists, we all do naive science. In June 2002 I was in the prep room of a school's science department. It was just before the afternoon lessons, and the teachers were all deeply involved with watching the soccer World Cup on TV. Murphy's Law is very clear about football on TV.

If you blink, they score; if you go out for a pee they score three times; if it's a penalty shoot-out the TV breaks

It was a penalty shoot-out, so I knew the TV would break, though I couldn't know whenabouts. A teacher came into the room, moaning about a student. He threw his book down on the table. The TV broke. All the science teachers turned on him: 'What did you do?' they spluttered. They examined the book, checked the table and looked all around that corner of the room. There was no connection, of course, but they were naive science teachers. Nobody is free of the instinct.

We all like simple answers. We don't like hard jigsaws. In our mind we would like to take a saw to all the bits that stick out, or soak the pieces until they're mushy, then squeeze them in the space we've made for them. In the next chapter we find just the tool for doing this, something to boil it, stir it and dye it any shade we like.

5

Emotions

When we start dealing with the emotions, the analogy with the jigsaw breaks down. Doing a wooden jigsaw is a cool process; it has nothing to do with emotion. Doing life's jigsaw is anything but cool. Everything in life has an emotional flavour. Mediaeval philosophers reckoned on four basic emotions, or 'humours': phlegmatic (easygoing), choleric (aggressive), sanguine (whimsical) and melancholic (depressive). (Clearly 'humour' has changed its meaning in the intervening years.) Some modern psychologists have proposed six, some eight, different flavours; these are based on recognition of facial expressions; anger and disgust, anticipation and joy, sadness and surprise, acceptance and fear. But there may be more quasi-emotions lurking beneath the surface; subtle moderators of mood which cannot be described as moods in themselves.

What is clear is that there is nothing out there, however trivial, without at least a smear of emotion, in the same way that there is nothing without a tincture of colour, a whiff of smell or a hint of sound.

Although we are born with many emotional responses already in position − instinctive − most of them have to be learned. For instance, the colour red is just a colour. If a child sees a man waving a red flag it thinks 'man + flag + red', and not much more. Later on it learns to attach emotions to that colour. A man waving a red flag on an army shooting range brings out a different emotional flavour (danger) to a man waving a red flag in the middle of the Chinese New Year celebrations (joy), or outside a presidential palace (revolution). The addition of emotions means that 'red' now signifies a lot more than it did.

Although they have been kept out of the picture up to now, emotions are at the base of all of Murphy's Laws, every single one. When Murphy's Law strikes – as the toast heads unerringly towards the floor, butter-side down yet again – do we feel indifference? Never! Anger, yes; gloom, perhaps; self-pity, frustration, depression, misery, guilt, irony, shame, sorrow, disgust, surprise, suspicion, contempt, anticipation, fear, remorse, awe, dismay, cynicism, disappointment, embarrassment, fury, shock or amazement, possibly; but indifference, no.

FEAR

The seat of the emotions is the amygdala, a small organ found in a lower level of the brain, called the 'mammalian brain' because it evolved in mammals long before the modern brain developed. The amygdala deals with the most important emotion – fear. Fear can be a life-saver. The rush of adrenaline, the pounding of the heart, the staring eyes, all the outward signs of fear stem from an inner preparation for fight or flight – an inbuilt survival mechanism as important as the reflex that jerks your hand away from a flame. Like that reflex, fear is automatic and swift, and like a reflex it will often eclipse other brain functions.

Once the amygdala identifies a threat it sparks off a cascade of hormonal responses whose main result is the appearance of adrenaline (epinephrine) in the blood stream, and the adrenaline sets you off on a panic, which we see during one of our more embarrassing Murphy experiences, in the bathroom.

The spider in the bath gets bigger every day

About 10 per cent of Europeans are arachnophobic. Yet 100 per cent of Europeans know that European spiders are harmless. So why the panic? Why, as soon as you set eyes on the spider, do you run screaming from the bathroom as if from the jaws of death? Can you not hear the little fellow calling 'Help! Help! Poor me!' Why doesn't your mind say, 'Hold on! There's no danger, just a poor little spidey who's fallen in the bath and can't climb out again!'

When you see the spider, the information from the visual cortex goes first to a relay station just above the amygdala, the thalamus. From there it splits and takes two separate routes; one goes to the cortex in the normal way, for thinking about, and perhaps doing something about sometime in the future. But the other runs straight to the amygdala, which takes one look and presses the panic button. Before you know it, you are outside the bathroom door. And that 'before you know it' is literally true. An fMRI brain scan (see p. 238) shows that the cortex, the bit that might do something about it sometime, is left out of the loop entirely. It only ambles along later, looks at the hormonal mayhem and records the 'emotion' of fear, at which point 'you' become conscious of the situation. This is the strangest thing; the last person to know about your panic is you. Everyone around knows what you're up to long before you become aware yourself. Although you tell everyone afterwards that you felt this strange sense of terror and ran from the bathroom, the truth is the reverse; you ran from the bathroom first, then you felt the terror.

The rapid action of the amygdala is necessary, because speed is essential in an emergency. This is not the time to engage the intellect – that spider could pounce at any second.

Which brings us to the Murphy moment – the central question: What emergency? Why pick on spiders? They aren't the least bit dangerous, at least not the ones we find in the northern European home. Nobody knows the answer to this, but it is possible that we are encountering a deep instinct here; a vestigial reflex dating back hundreds of thousands of years, before *Homo sapiens* had travelled north from Africa, where some spiders do bite and some are poisonous.

It is already known that quite complex fear responses are inherited. Horses have a fear of snakes.

Without ever having seen a snake before, they will shy away from the sight of one. Even seeing a hosepipe will set them off. Our fear of spiders is just as automatic, untrained and resistant to any attempt to shake it off.

Some researchers say we are reacting to the way spiders move. Like panthers, spiders are hunters, and like panthers they lurk, stalk, creep, pounce on and eat other animals. Maybe we fear we might be next on the menu. Whatever the reason, it is a primeval response, inappropriate for a suburban bathroom. This phobia must join goose bumps and the appendix as a piece of evolutionary flotsam.

(While we're gazing in the bath, here's another thing; why do so many of us believe that the spider came up the plughole? The naive scientist in us (p. 67) has an explanation, based on the simple notion that things that sit together belong together. The spider sits next to the plughole, therefore the spider must have come up the plughole. QED. Friends may point out that there is a U-bend on the outlet – the spider would drown trying to swim round it. Naive scientist says, 'he must have an aqualung'.)

Fear of one sort or another is our constant companion:

When you don't have any money, the problem is food.

When you have money, it's sex.

When you have both it's health.

If everything is simply jake, you're frightened of death

(J.P. Donleavy, b. 1926, Irish author)

Most of us, like Donleavy, have a continuous, low-level hum of fear. The fear has no focus; it's just a nebulous feeling of unease. The cortex automatically looks for something to attach it to. Naive scientist as it is, it will attach the anxiety to whatever is on the mind at the time, be it food, sex, health or death.

This is a bit topsy-turvy; we naturally believe that the emotions are instructed by the mind, not the other way round.

In 1962 Schachter and Wheeler tested if this is true in an experiment[1]; participants were injected with either adrenaline or chlorpromazine (which inhibits arousal), then asked to watch a slapstick film. The adrenaline group laughed hilariously, the chlorpromazine group sat beside them, stoney faced. Afterwards the adrenaline group said they thought the film was hilarious, but the chlorpromazine group couldn't think what all the fuss was about. Their view of the film was dictated by the state of their hormones. So the mind does not instruct the hormones; it is happy to be instructed by them.

All news is bad news

The most regular division of emotions gives six categories: surprise, happiness, disgust, fear, anger and sadness. Four out of six of those are unpleasant, so it's not surprising that most of our life is concerned with unpleasant things. Your brain has plenty of fears to keep itself busy without wasting time on things that are going OK. It concentrates on the problems. The TV news is mostly about disasters. If it told of things that were going right in the world it would be perhaps more pleasant, but also a very, very, very long programme.

The best way to be late is to give yourself plenty of time

Without an emotional impulse a job will tend to languish at the bottom of the 'things to do – urgent' list. Whatever the cortex may mutter about giving oneself enough time to do the job and avoid last-minute hitches, there it will sit until the week before it is due, when you are gripped at last by panic. What can be better than a deadline to get the amygdala to crack its whip? A well-trained amygdala will scoop up memories of dole queues, loss of friends, eviction and starvation, and keep you and your cortex fired up all night every night for a week. And then you find out how true this is:

The first 90 per cent of a job takes up
the first 90 per cent of the time. The
last 10 per cent of the job takes up
the second 90 per cent of the time

(Arthur Bloch, Murphy's Law, 1977)

LOERVE

Murphy loves Love; it makes fools of us all. Sex is such an imperative; a male moth will fly for miles and search for hours if just one molecule of female pheromone touches his antennae. But the act of sex is not inevitable, not with insects and not with humans. Females from the fruit fly to the bar fly tend to be selective, putting the male through tortures, mating dances and courtship rituals before allowing him to copulate.

When the penis stands up, the brain flies out of the window

(Chinese proverb)

What actually happens to your head? In 2001 Professor Semir Zeki and Andreas Bartels[2] sought out volunteers who were 'truly and madly' in love. The volunteers had their brains scanned as they looked at a picture of their lover, and the results showed just what goes on up there in our brain as we fall in love. The medial insula, which is a highly connected region with links to all the sensory areas of the brain, becomes very busy, as does the anterior cingulate, which is known to respond to euphoria-inducing drugs. Pleasure centres in the deeper and more primitive basal ganglia region of the brain are also active. Reduced activity in the postcingulate cortex increases a sense of euphoria. Hormones in the recipe? Dopamine, phenylethylamine, testosterone, serotonin, oestrogen, oxytocin, various pheromones . . . It's not surprising that love makes a kind of madness happen. All this, just to get the next generation started.

Most women select their husbands in lighting they would not choose a car by

(Anon, twentieth-century USA)

Once having made up your mind, you don't want to un-make it. In the emotionally charged areas of love, a picture, once painted, should not be tainted by reality. Not only does your mind see what it wants to see; it actively avoids seeing what it doesn't want to see. Love is blind; actually love is deliberately blinded. The lighting is set low, and plenty of sultry music, perfumes, exotic foods and interesting drinks are deployed.

Love looks not with the eyes, but with the mind, and therefore is winged Cupid painted blind

(William Shakespeare, 1564–1616; *A Midsummer Night's Dream*, 1595)

When she sees that look of love in his eyes, what is she actually seeing? I'm afraid the look of love in a gloomy restaurant can be a bit of an illusion. One of the effects of passion is to dilate the pupils. The 'look of love' is the look of big dark pupils (as any painter will tell you). In a

darkened restaurant the pupils adjust to the gloom by dilating. Bingo! The look of love at the flick of a switch, literally.

What if the beholder does not have their love returned by the adored one? How to cure the pangs of unrequited love? Wendy Cope has the answer in her poem, 'Two cures for love':

1. Don't see him. Don't phone or write a letter.

2. The easy way: get to know him better.

The heart has its reasons that reason knows nothing of

(Blaise Pascal, 1623–1662, French mathematician and philosopher; *Pensées*, 1670)

For those who are willing to volunteer for the roller-coaster ride of love, a surprisingly effective alternative to the soothing restaurant is a real roller-coaster. Anecdotal evidence says that affection can blossom on a white-knuckle ride.

'Love keeps you young', somebody who was old once said. The sayings below are forever young, because however much we nod sagely at them and say 'yeah, how true', and 'amen', we will go back and back and re-offend as soon and as often as we can.

- A woman without a man is like a goldfish without a bicycle (feminist slogan, 1970s)

- A man without a woman is like a neck without a pain (bumper sticker, 1980s)

- Men – can't live with 'em, can't live without 'em (anon)

- A second marriage is the triumph of hope over experience (Samuel Johnson, 1709–1784, English author, critic and lexicographer)

- A bachelor is someone who never makes the same mistake once (anon)

- A successful marriage requires falling in love many times, always with the same person (Mignon McLaughlin, b.1915, American journalist)

Is this another example of the mind misreading the emotions?

In 1974 Dutton and Aron[3] showed the illusion that can happen when the senses are a little aroused. The Capilano Canyon in British Columbia is a tourist spot famous for its 200 metre (220 yard) rope bridge. One day a sexy-looking female researcher was to be found going up to male visitors to the canyon and asking them questions, supposedly about the effects of scenery on creativity.

Afterwards they were asked what they thought of the interviewer. Some of them had been approached by her on the wobbly rope bridge. They were in an 'aroused' state of course, because they were scared, but this seemingly affected their opinion of the girl: the men on the bridge thought the girl more attractive than did the other men, who apparently had their feet more firmly on the ground in more ways than one. So the senses can be fooled, it seems. The thrill of fear can be mistaken for the thrill of love.

So here you are, adoring, worshipping your icon. He/she is undoubtedly perfect. But how about you; are you perfect enough for your partner? Have a look in the mirror . . . Ugh! What is that staring back at you? Surely you weren't that fat yesterday . . .

The past few chapters should have prepared you for this moment in front of the mirror. Here is confirming proof that we see what we want to see, and we have no idea what's actually there. Sometimes we can look in the mirror and there is a svelte, dashing, gorgeous thing looking back, sometimes it's the creature from the black lagoon staring at us. It depends on how our emotions are running on that day.

And while we are casting our hypercritical eye over the shape reflected back at us, we can forget our intelligence, wit, humanity and generosity. There are more important things to deal with. For Her it's . . .

On the evening of the big date a pimple
appears in the middle of your nose

And for Him it's . . .

However hard you comb it, one
hair always sticks up

We become obsessed with tiny details, details which our beloved will never notice, because they are too obsessed with their own tiny details. He is convinced he spotted a grey hair, while she is desperate to get rid of those wrinkles round the eyes, slapping on the age-defying cream. For goodness sake! They're only 17 years old!

ANGER

Passions make good servants and bad masters

(Roger L'Estrange, *Aesop's Fables*, 1692)

The amygdala gets its name from the Latin for almond, which it resembles in more than just its shape – sometimes it can make you go nuts. Anyone who has been

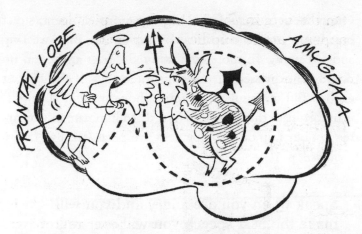

really off-the-scale angry (and that would be all of us), knows just how nuts it can make you. You can meet your ancestors, the apes, as they express themselves through you in a display of pure simian aggression. On the other hand, the more rational frontal lobes can bring you back. The connection between the two is important. When something gets you hot under the collar, the frontal lobe can tell the amygdala to cool down. Compared to the amygdala, the frontal lobes are a much more recent evolutionary development, only one and a half million years old. Their moderating influence on the ancient amygdala is what makes our modern environment – cities, democracy, living on top of each other, grinning and bearing it – possible.

Anger makes dull men witty, but it keeps them poor

(Francis Bacon, 1561–1626, Elizabethan scientist and courtier)

Anger expresses itself vocally. Chimps in the jungle emit high screeches when roused, as a warning to their enemies.

Often modern man's rages are no more eloquent; street brawls are not accompanied by rapier wit, just swearing and screaming. In more sophisticated circles anger can no doubt be expressed in poetic pyrotechnics of the highest calibre. But an insult is an insult, however scintillating. It is unfortunate that the same fires that illuminate our repartee can burn up our employment prospects (see also *l'esprit d'escalier*, p. 101).

Speak when you are angry and you will make the best speech you will ever regret

(Ambrose Bierce, 1842–1914, American author and critic)

The amygdala, hot with passion and ready to fight, is connected to the frontal lobes, which reckon that to win the war you may have to lose the occasional battle. The words of ancient sages on moderation speak on behalf of the frontal lobes. In Korea they say that if you kick a stone in anger, you hurt your own foot. Buddha said, 'Feeling anger is like grasping a hot coal with the intent of throwing it at someone else; you are the one who gets burned'. In Germany they say, 'write an angry letter – but don't send it'.

Too often, though, we 'see red', gripped by a force beyond our control. The daze that descends after doing something very brave or very stupid ('What happened? Where am I? Everything went blank') demonstrates the amygdala in full spate. The amygdala's purpose is to save your life in an emergency. The forces it has at its command are formidable, and its authority is total. It will prepare you for strenuous action, numb all pain, alert your senses and speed up your brain's processing power, so you can make multiple decisions more rapidly than ever before (so rapidly that the world outside will appear to slow down during your crisis). But those decisions will be suited to a Palaeolithic world rather than the more measured, urban, urbane even, expectations of nowadays. Crimes of passion can result. The plea of 'diminished responsibility' is there in law to explain the horrors that can happen when the frontal lobes lose control.

It is hard to get the right balance. On one hand we admire bravery. Jean Paul Richter said, 'A timid man is frightened before a danger, a coward during it, a courageous man afterwards'.

But the same quickness of reactions which produces heroic acts can also lead to tragic acts. Brain scans of 41 convicted murderers[4] showed reduced frontal lobe activity. These people murdered because their frontal lobes couldn't control their amygdalas. By the time the frontal lobes were activated the damage was done.

Patience is something you admire in the driver behind you and scorn in the one ahead

(Mac McCleary)

Road rage, supermarket aisle rage, all the rages that get the hairs prickling on the back of the neck, these are the Palaeolithic heritage breaking through. The powers of control of frontal lobes over amygdala need to be trained through socializing. For most of us, this socializing is nine tenths of our family and school life. At home we learn to wait for rewards and use reason rather than tears to get what we want. At school, respect for each other and co-operation with the teachers is so vital that it never even needs to be mentioned in the core curriculum.

There are many examples of what happens when the training is left out. Children rescued from the horrific orphanages in Romania in the 1980s acted as if they had no emotional control at all. Their environment had been dehumanized from birth; they were virtually abandoned, given no more than basic food and shelter. The path from frontal lobes to amygdala had never been trod. Danya Glaser[5], studying orphans who had been adopted into American families, notes that though they slowly improved their social skills, the longer they had been institutionalized, the longer they took to control their emotions, which could swing from love to hatred in seconds. With emotional control, as with everything, it's 'use it or lose it'.

The words a man can never utter and live: 'might you possibly be a little premenstrual?'

Premenstrual syndrome (PMS) happens because the body's hormones move the emotional goalposts, leaving the brain offside. Changes happen to the balance of serotonin, low levels of which have been linked to depression; gamma-aminobutyric acid (GABA), which is associated with anxiety; and of endorphins, neurochemicals that alter the pain threshold. In brief, the emotions are all this way and that, and the mind is often unaware of the fact. If the frontal lobes are not able to adjust to the new state of the emotions, they can trip behaviour into a state of fury quite out of the blue.

PMS has been used as a mitigating circumstance in several court cases, including a significant trial in 1980[6]. By that year Sandie Craddock had already had several run-ins with the law. She had been up for theft, arson and assault. But in 1980 she killed a co-worker, so now she was on trial for murder. The police discovered that she had kept a diary, and at the trial a doctor noted that her acts of violence happened every 29 days. She was found guilty of manslaughter based on a plea of diminished responsibility caused by PMS.

You think of the right thing to say just after you shut the door

Called '*l'esprit d'escalier*' ('the spirit of the staircase') from the old experience that it is as you clump down the staircase with the sound of the slammed door ringing in your ears that you compose the brilliant, elegant and stunning put-down which would have settled the argument if only you had thought of it ten seconds earlier.

If you had been more ruled by your amygdala just now you would have behaved like your earliest ancestors; screamed, bitten and flung furniture. It would be clear who had won the argument – whoever was left alive. If, on the other hand, you ignored the amygdala entirely you would have sounded flat and disinterested. The task is to balance 'war-war' against 'jaw-jaw'.

The ability to win an argument nowadays has nothing to do with violence, but also not too much to do with truth, justice or reasoned discourse. It has plenty to do with practice at arguing. It is a craft, like playing the piano. Rhetorical devices, trills and flourishes must be learned and rehearsed many times before you will be ready to knock them dead.

This is best seen in its most sophisticated guise, in the courtroom. Lawyers, to be fair to their trade, are hired to win the case, not to reach a just solution. Their rhetorical skills are delivered with measured amounts of passion, the amygdala held on a tight leash.

It's a difficult balance. Without passion, fairness and balance have a chance to prevail. Yet without passion there would be no desire for fairness in the first place.

If parents shout at you, it's because they're wrong.

If they don't shout at you, it's because you're right

(Why the World is Wonky)

Hire a teenager, while they know everything

(Sign in chip shop, Banbury, England)

As the child rockets up and beautifies into an adult, the parents shrink and uglify into fuddy-duddies. Physically, as the youth grows in height the parents appear to shrink. Equally dramatic but less obvious changes are also under way in the pubescent brain. The hormonal landscape is in upheaval and desperately needs the frontal lobes to keep it under control.

And here there is a big problem. fMRI scans (p. 238) of developing brains have shown that maturation takes longer than we thought. The focus of the studies[7] was not on the nerves themselves, but their insulation. Between the ages of 14 and 25 a layer of fatty myelin progressively envelopes all the neurons in the brain, making them more efficient, just as insulation on electric wires improves their conductivity. The last areas to be myelinated are the frontal lobes, in the years up to age 25. So young adults are right to feel anxious about their reactions. Testosterone and oestrogen are swirling all around, but the frontal lobes are not yet ready to anchor the raging hormones. So it is that youths are grown up and mature in their passions, but childlike in their ability to control them.

Mark Twain reflected on his upbringing; 'When I was a boy of 14 my father was so ignorant I could hardly bear to have the old man around. But when I got to be 21 I was astonished at what he had learned in 7 years'.

AVOIDING EMOTION

People neither mean what they say nor say what they mean

A careful use of language can help to remove the emotional flavour from a confrontation by avoiding upsetting the amygdala. In the Houses of Parliament, for instance, there are strict rules about the use of inflammatory words like 'cad', 'stool pigeon', 'guttersnipe', 'murderer', 'swine', or 'cheeky young pup', all of which, according to Hansard, will cause the honourable members a fit of apoplexy.

Journalists who are obliged to write something nice about somebody who they can't stand will find words to convey the right emotion to the right ears, while avoiding distressing the wrong ones. John Leo, in *Journalism for the Lay Reader*, points us to the true meaning of 'soft-spoken' (mousey), 'loyal' (dumb), 'high-minded' (inept), 'hardworking' (plodding), 'self-made' (crooked) and 'pragmatic' (totally immoral).

Words can take on a meaning they didn't intend. A 'cretin' is a fool. Well, no it isn't, it was originally used to describe a medical condition, and has been corrupted by popular usage. Out of respect to the people it was originally used to describe, they are now called 'congenitally hypothyroid'. It is to be hoped that 'cretin' will fall rapidly out of use.

In 1866 Dr John Down described a group of retarded children in a mental asylum who, it seemed to him, greatly resembled Mongolians. This neatly fitted into the English notion that Easterners were backward, so the term 'mongoloid' was used to describe them. One hundred years later it slowly dawned on doctors that this rubric was not kind to Mongolians and was deeply offensive to the sufferers, and so the more respectful 'Down's syndrome' was adopted.

In the face of a disaster, careful phrasing is counter-productive. Engineers who have to write 'failure mode effect analysis reports', in which they list everything that can go wrong and what will happen if it does, ought to be open and candid, but they are inclined to pussyfoot, saying:

'uncontrolled thermal event' when they mean 'fire'
'unplanned loss of containment' when they mean 'bursting'
'spontaneous rapid disassembly' when they mean 'explosion'
'deconstructive deceleration' when they mean 'crash'

With each new military adventure, we are used to hearing sanitized phrases such as 'collateral damage', meaning civilian deaths, 'ethnic cleansing', meaning genocide, or 'friendly fire', which is exactly not what it says.

The death of one person is a tragedy; the death of 5,000 people is a statistic

(Josef Stalin, 1879–1953,
Soviet dictator)

Stalin sums up our problem. Modern humans have to deal with big numbers, while our primitive forebears had a three-number counting system; 'one – two – lots' (see p. 181). For Stalin (a man very much in touch with his primitive side) anything over two was too big to grasp, and therefore too small to get emotional about. We can gasp at Stalin's cold-blooded cynicism, but we must not forget how hard it is for us all to care about human tragedy. The amount of compassion we feel doesn't go up step by step with the scale of the tragedy, but seems rather to fade away rapidly. From street beggars to famine relief, repeated requests for generosity lead to 'compassion fatigue'. Charities have to spend a lot of money on new advertising campaigns to keep the donations coming in.

Knocking Stalin is easy – he engineered the deaths of 20 million or more of his own people between 1924 and 1953 – but the rest of us make our contributions to the world's problems. The West's demand for wood leads to deforestation of poor countries in the East. Chopping down the trees destabilizes the soil, which leaves it vulnerable to floods and avalanches, which are the cause of many of the human disasters we see on TV. At the end of the disasters, after the western camera crews have left, the people find themselves with no topsoil, and therefore no agriculture. But it all happens so very far away . . .

The further away the disaster occurs, the greater the number of dead and injured required for it to become a story

(Fuller's Law of Journalism)

Brains of the future, if they evolve like crazy, will be able to think globally – to realize the effect that action in one country has on people in another. For the time being our technical brilliance is running ahead of our emotional maturity, and we are stuck in the uncomfortable position of being able to act globally, but think only locally.

Time heals all wounds

(Geoffrey Chaucer, 1343–1400, English author;
Troilus and Criseyde, 1385)

Another way to remove emotion from our lives is to let the passage of time do its job. Forgetting is as vital to our emotional lives as remembering. If all the dreads, terrors and fears of our lives remained as powerful whenever we recalled them as they were when we first had them, we would dissolve in adrenaline. Unfortunately some wounds will not heal. Memories of traumatic events can continue to haunt for decades. Scientists are looking for ways to help the healing process.

In 2002 Beat Lutz, at the Max Planck Institute of Psychiatry in Munich, identified the neuroreceptor CB1, a chemical related to cannabis, as the molecule which helps us to forget past traumas[8]. So far the Munich studies have been confined to rats, but scientists are interested in developing the research further. It might, they say, lead to a treatment for post-traumatic stress disorder (PTSD), the inability to extinguish scaring memories found in victims of tragedy or in combat troops.

A look around would suggest that the public have already done the research – the rise in the use of cannabis in America which occurred after the start of the Vietnam War in the mid-1960s, particularly among Vietnam veterans, suggests that many people were ahead of the game.

The advertising industry attempts every trick to play on your emotions. If it succeeds, it's because it has managed to coax an emotion out of you – love, of sorts, is the emotion estate agents plug for, fear is preferred by insurance salesmen, anger is encouraged by political parties.

Everything looks useful until you buy it

How did that useless, shapeless, valueless, ugly thing ever find its way into your living room? You bought it? How could you have been so stupid!?

This is how; the advertising industry targets the naive scientist in us all by putting the thing it wants to sell next to something that encourages an emotion, be it pride, power or love. When two things happen together they are linked together. Pavlov pointed out this simple fact 100 years ago with his work on dogs and the conditioned reflex. You feed a dog; you ring a bell at the same time. After a few goes, when you ring the bell the dog looks around for food. The same works for humans. You drape a bikini-clad model over a car, the human buys the car, then spends the next two days looking through the glove compartments for the girl. Associate two emotions together and the response is automatic. Celebrity endorsements, product placement, reassuring voice-overs. After a 5-minute exposure you are hooked. Only after your emotions have faded a little do you realize just what you have been lumbered with.

OBJECT-IVITY

A lot of Murphy's Laws involve us shouting impotently not at people, but at things. We have an extraordinary gift for animism, attributing personality to inanimate objects. And with the personality comes emotions, intentions, motivations. In 1944 two experimenters, Heider and

Simmel, produced a film showing two triangles and a circle moving into, out of, and round about a square[9]. People watching the film worked out elaborate stories of love and rivalry between the triangles, vying for the circle's affections.

We know animism is inherent in us. Jean Piaget, the child psychologist, recorded many cases of animism in children – smacking the table that bumped into them, telling a stone not to harm the garden by dropping on to it, asking a plank not to cry when a hole is drilled into it. That instinct should be trained out of us when we grow up. But no, it's still there. We refer to ships as 'she'; death is 'he'. Death is not a neutral event. Death has purpose: 'he comes for us'. We think things are active agents, working for us or against us. And mostly against us.

Animism walks alongside naive science. Together they are the brain's answer to Laurel and Hardy.

Never let a computer know how much hurry you are in

Animism makes us attribute animal properties – even human properties – to everything, including evil computers who can tell the very instant you are going to press the 'save' button, and crash nimbly into oblivion a microsecond ahead of you. As Scott Adams, creator of Dilbert, points out, technological marvels such as computers were created by geniuses, then dumped on idiots like you and me to struggle with.

Post-Palaeolithic man, with his naive-science, pre-programmed, animistic brain, can only perform a ritual dance around the genius creation which has just collapsed for no reason, screaming 'What!!! How can you DOOOO that!!!! Please undo!! Pleeeeease!! I'll give you many succulent pigs if you un-crash just this once . . .'

Objects behave better when you shout at them

Swearing really works. Put on a good display for the benefit of the spanner, the bolt and the bike brakes and they'll slip together in no time.

The psychology of swearing is unknown. It is spoken of as a releaser of tension, but how, where or why is clearly dying for a research project to be launched at it. Like a sneeze clears the sinuses, an oath sweeps the cortex and leaves the mind better able to deal with the task. However, it is a mistake to say that swearing caused the bike parts to pull themselves together, for then we are slipping into animism. That might lead us to develop unhealthy beliefs about all sorts of things.

If you decide to walk on to the next bus stop, the bus will turn up when you are midway between stops

This isn't the place to talk about how innocent of sin is the bus. (You can find out why buses go round in threes on p. 192.) Here we expose the cussed nature of the beasts. With low cunning they skulk behind the post boxes waiting, waiting, until you are precisely midway between stops, then with a joyful whoop they streak past and on over the horizon. The animist in us gives the bus free will and the naive scientist in us deduces malice was the motive.

Chapter 8, on science, will deal with the real lives of inanimate objects (or rather, the real lack of life); the toast performs a delicate Fosbury flop to land on the floor butter-side down; socks in their drawer take a secret ballot to see who's going to go missing; when you want to write down a phone number the ink sprints out of the pen, which rockets away from the notebook, which vaporizes as you reach for it; the missing left glove is quietly laughing at you from the back of the sofa while you search for it in the cupboard, and so on . . .

What is the point of animism? The little cherubs on ancient maps who blow into the sails of schooners show it. Worshippers of Nature and tree-huggers show it. Basil Fawlty shows it when his car stalls; he beats it with a tree branch as punishment. The old sailors prayed for good weather, as if it would make a difference to the great boiling, chaotic weather systems. They might just as well have sent £50 to the butterfly in Patagonia, asking it not to flap its wings and create a storm in the North Atlantic. (See the weather forecast on p. 190.) Gamblers ask Luck to 'be a lady tonight', hoping the cards will shuffle themselves into the right order.

Animism is counter-productive. It's pointless. It clearly confers no evolutionary advantage. I believe that animism, like arachnophobia, belongs in an evolutionary backwater. It is an overspill from the development of human empathy, (see mirror neurons, p. 117). Our ability to read the minds of other people spills over into reading the minds of objects, even when there are no minds to be read.

Inanimate objects move just fast enough to get in your way

As soon as you mention something; if it's good it goes away; if it's bad it happens

The last screw won't undo

Whatever you buy, the day after the guarantee runs out, it snaps

Nature always sides with the hidden flaw

Any situation demanding your undivided attention occurs simultaneously with a compelling distraction

Inside every small problem is a big problem struggling to get out

The chief cause of problems is solutions

Archimedes updated – When a body is immersed in water, the doorbell rings

Gerrold's Laws of Infernal Dynamics:

- An object in motion will always be heading in the wrong direction.

- An object at rest will always be in the wrong place.

(All from Arthur Bloch, *Murphy's Law*, 1977)

SUMMARY

We aren't going to evolve away from our emotions. The next natural step for mankind is not *Star Trek*'s Mr Spock, because an emotion-free life is utterly impossible. Antonio Damasio[10] tells of a patient, Elliot, who had a dangerous tumour removed from an area in the front of his brain. Unfortunately the operation severed some connections between his frontal lobes and his amygdala. In many ways Elliot made a full recovery. His memory was intact. His body functioned normally. He could do most things. His problem was that he couldn't make up his mind which things to do. Because nothing felt any more important that anything else, Elliot could spend all day vacillating, unable to make the simplest decision. Without that waft of emotion, however slight, that the amygdala could impart, he was rudderless. The amygdala guides our preferences, which guide our life.

The case of Elliot shows up how important emotions are to normal life. But we also see that a 'normal' day includes quite a lot of 'weird'; feeling terror in the company of harmless spiders, falling in love with the wrong person, falling in hate with a piece of useless technology. It makes you wonder if emotions are overrated.

For the purest of pure science, emotions are an enormous stumbling block on the path to truth. Giving human motivations to inanimate objects stops us from seeing how they really work. The greatest scientists, it seems, were and are the ones who look for cold facts. Henry Cavendish, a brilliant English scientist of the seventeenth century, who made discoveries that were 100 years ahead of his time, wore clothes that were 100 years behind his time. He never spoke above a whisper and shunned human company. He was really not interested in humans at all. His extraordinary scientific insights were possible because he avoided getting emotionally involved in the inanimate world. But this was offset by his tendency to do the same with humans. Newton and Einstein showed similar traits. Their discoveries were possible because they were outside the emotional stream, able to see things more coolly.

We are wary of that coolness. The cold scientist is somehow a dangerous scientist. Like Frankenstein or Edward Teller, the father of the H-bomb, it seems, they do things because

The five different kinds of scientist: The Saviour of Humanity, Trust Me, they can be done, not for any humane purpose. The ability to do wonderful things with science needs to be constrained by an emotional understanding of the consequences.

In rare cases of tragic accidents where the amygdala has been completely severed we can see how essential the emotions are. These people lose all interest in the outside world. One object feels exactly like another – and humans are included as 'objects' – because they evoke no emotional response. They feel no happiness or sorrow, they are emotionally flat. They lose all interest in life, since life and death are equally stripped of meaning.

Without emotions we are no more human than an amoeba. Even Mr Spock, the supposedly emotion-free Vulcan in *Star Trek*, has to have emotions – and quite a lot of them – otherwise what could have motivated him to 'boldly go' week after week? But without control of the emotions, we will be in danger of going not only boldly, but also recklessly, and Murphy will be right behind.

The Evil Genius, Bean-counter, Absent-minded Prof . . .

6

Public Opinion

Do we think we've completed the jigsaw? We've collected the pieces, put them in place and filled in the gaps, but it's not over yet. We haven't asked everyone else's opinion. In spite of all that you might think from what has gone before, your own picture counts for nothing compared to that of the crowd around you.

Do you know of people who are perfectly docile at home but turn into wild party animals on Fridays, or who are uncontrollable at home but sweet angels in company? You know what I'm talking about, then. This chapter looks at what happens when we are in groups. Whatever we believe our 'core' inner self to be, we become a different person at parties, on committees, with children, at political demonstrations, at football matches, even with just one other person in the room. We have a different 'me' for all of them. When we get into groups we lose our individuality. As we will find out in this chapter, we make decisions in groups which nobody in their

It is hard to go against the flow.

right mind would make on their own. Murphy's Laws are the only laws.

MIRROR NEURONS

The basic currency for group behaviour is the mirror neuron. V.S. Ramachandran has predicted that 'mirror neurons will do for psychology what DNA did for biology'. The phenomenon of mirror neurons is real, measurable and remarkable to behold.

In 1998 Giacomo Rizzolatti, using fMRIs on monkeys, discovered that when one monkey watches another monkey trying to solve a problem, his brain lights up in exactly the same way as the other, it mirrors the other's thinking almost neuron for neuron[1]. The implications of this are staggering. The two brains are slaved to each other. The learning of monkey X is transferred to the brain of monkey Y instantaneously, without any communication between them. No chat, just zap, across that gap. Merely by watching monkey X, monkey Y is now free to do the same as he has seen monkey X do, or perhaps do it slightly differently and test what happens, and monkey X can watch, and learn, and think, and try another way, and this way they can solve the problem. Two heads are decidedly better than one when mirror neurons are involved.

We shouldn't be surprised at the mind-reading skills of those monkeys. After all, their brains are all built from the same materials and to the same basic design. Since they're wired the same they fire the same. It should be comparatively easy for one to 'hack' into another's brain and feel perfectly at home there. Human brains are similar in structure to monkeys. We love watching monkeys in the zoo because we understand what they're up to; we can read their minds. Can we read dolphins' minds? Can we read dogs' minds? Can dogs read our minds? Is that why they can tell when we're going walkies? Are we all one big happy, empathizing, parallel-programmed mirror of each other?

We certainly use mirror neurons amongst ourselves. Call it 'empathy' if you like, but when we wince watching someone else have an injection we are feeling their pain. When we cry in the cinema at someone else's sorrow it's our mirror neurons which are making us do it. A jolly person in a group makes the whole group jolly. A misery makes the group jolly grumpy.

We can mirror each other from an early age. Andy Meltzoff[2] noticed that babies as young as 40 minutes old can mimic the face in front of them, pulling faces and poking the tongue out, for instance.

It is easy to make people's minds automatically mirror each other. Next time you are on a crowded bus or train, yawn: soon half the carriage will be yawning too. Nothing they could do, their brains are slaved to yours.

Mirror neurons are behind a major Murphy's Law:

Nothing works when people watch

You've been practising your violin piece for a fortnight. It went fine while you were on your own, but now, in public, when it really counts, with everyone's eyes on you, in the expectant silence, your brain melts. What has happened? Understanding mirror neurons helps to explain.

For a start, playing a musical instrument is not easy. Coordinated movements involve several areas in the brain. The cerebellum deals with the basics: Are you upright? Do you know where your fingers and thumbs are? The putamen stores the finger skills you already know, the hippocampus helps with the new fingerwork, the motor and sensory areas instruct the actual muscle movements, the frontal cortex thinks about interpretation. Your practice sessions are as much to do with coordinating these departments as about anything else. When they are functioning smoothly your fingers will fly, seemingly without you thinking about them. In fact it is essential that you don't think about them, you will interrupt their private dialogue. This is what is called being 'in the zone', although it would be better called being 'out of the zone', because you should keep out of the way and let the brain get on with its many tasks.

The arrival of an audience changes everything. Your cool is shattered by all sorts of new thoughts from unexpected parts of the brain; 'Is my hair straight? I hope nobody has noticed the pimple! Mustn't forget to bow! Have I **been** yet?!'

In addition, the presence of others in the room causes your mirror neurons to hum: Your audience will appreciate the music 'through' you. But also you start listening to the piece through the audience's ears. Your practice sessions did not prepare you for these neuronal ripples, and of course as soon as you become self-conscious your coordination goes. You have to learn to cope with the strange new thoughts to help your brain get back 'in the zone'.

The simplest cure for shyness is to do your practice in company. In the same way that walking along cliff paths makes you more at ease with cliffs, practising with other people in the room puts you at ease with an audience.

MEMES

Thirty years before the discovery of mirror neurons, Richard Dawkins foreshadowed them in his 'meme' theory. The author of *The Selfish Gene* proposed in the same book that DNA is not the only thing that is transmitted down the generations: learned characteristics, cultural units, which he called 'memes', also have a life of their own. Like genes they are copied and they mutate; useful ones survive and spread, useless ones fade away.

Meme theory explained the remarkable discovery made on the island of Koshima in 1952[3]. In order to study the

Japanese monkey, *Macaca fuscata*, scientists were luring them out of the forest with sweet potatoes dropped on the beach. The monkeys liked the taste of the sweet potatoes, but not the sand. An 18-month-old female named Imo found she could solve the problem by washing the potatoes in a nearby stream. She taught this trick to her mother. Her playmates also watched and learned, and they taught their mothers. Gradually all the monkeys learned the art, and generations later they still do it. When a young monkey sees an older one wash the potatoes, her neurons mirror the other's and there it is, instantly embedded.

For us humans the range of memes is wide, far beyond the range of potato-washing. Anything which is learned and passed on to the next generation counts. On the small scale, a catch phrase is a meme; 'bear with me' is a meme. On a grander scale, social customs, national attitudes and religions are memes.

The first human to find a use for jagged flakes of flint was watched by his group, who copied him. The idea proved popular enough for a million-year-long slice of life on Earth to be named after his idea – the Stone Age. Fifteen thousand years ago another human or group of humans in Iraq tried out something more complicated; settling in the same spot instead of roaming the plains. They planted crops and corralled some animals, and agriculture was invented. Archaeologists can map the spread of agriculture from Mesopotamia westwards at a steady 1 km every year, as each community mimicked its neighbours, adopted the meme and passed it on.

What is the difference between a meme and a gene? Eating mushrooms comes under 'gene', since it's to do with the instinct of feeding. Knowing which mushroom to eat is a meme, developed over hundreds of years. Many people died finding out what mushrooms not to eat. This meme comes at a cost, and will be passed very carefully on to each new generation.

Working out which mushrooms not to eat.

Memes are as tenacious as if they were genetically programmed, and like genes they very often leave a trace long after they have mutated away. Part of our DNA still creates goosebumps on our arms even though another part has stopped providing decent hairs to go in them. That's a piece of gene junk. Buttons on jacket sleeves, a hangover from when sleeves could be rolled up, are meme junk.

Of course, to pass the meme around, you need a crowd. Fortunately we always manage to find one:

When you set out on holiday, so does
everyone else. When you set off home again,
so does everyone else. If you set off at 4 a.m.
to beat the rush, so does everyone else

Ironic that we're so irritated by the crush, but yet we
arrange all our celebrations to happen together. Some
think-tanks in the past have suggested staggering Christmas
and Easter so that everybody celebrated on different days,
avoiding the crowds. You can imagine how popular they
were. No – we all need to crowd in together. It is the
human way. We are gregarious animals. Sitting in a traffic
jam all Saturday afternoon because everyone else is trying
to get to the same shopping mall at the same time – we
love it. We love hating it. In the summer you can give us
a nice empty beach to ourselves, but sooner or later we'll
be running along the shoreline until we find a bay packed
with heaving blubber and squealing
kids, then we'll complain
deliriously about the lack of
privacy. From Christmas
carols to Cup Finals we
all cluster up together
at every opportunity.
And while we're
there we'll display
and share our
memes . . .

When people can do as they please they all do the same

(Eric Hoffer, 1902–1983, American author;
The Passionate State of Mind, 1955)

However fanciful our thoughts when we are alone, when we form groups our thoughts converge. In an imaginative experiment in 1937, Muzafer Sherif[4] showed this using the **autokinetic effect**, a visual illusion. If you sit in a darkened room with a single tiny point of light in it, the light will appear to move, because without any fixed frame of reference, the brain will be unable to stabilize the image. Sherif put several students in the room and asked them to watch the light. He didn't tell them the light was fixed. Each of them thought they could see it moving. For each of them the amount and direction of movement was different, of course, because it was a fantasy. But when Sherif invited them to talk about what the point was doing their reports began to converge. After a while all of them reported it moving the same way. In other words they coordinated their fantasies. Their minds flocked like birds.

Memes account for fashion in clothing. We could all dress completely and utterly differently, instead of which we check out what everyone else is wearing, and do the same. James Laver, in *Taste and Fashion* (1937) summed it up in his universal 'Law of Fashion'. This is how we all think:

Indecent	10 years before its time
Shameless	5 years before its time
Daring	1 year before its time

Smart	just right for now
Dowdy	1 year after its time
Hideous	10 years after its time
Ridiculous	20 years after its time
Amusing	30 years after its time
Quaint	50 years after its time
Charming	70 years after its time
Romantic	100 years after its time
Beautiful	150 years after its time

Straight hair must be curled; curly hair must be ironed

It is a racing certainty that by the time you reach the age of desperate conformity to fashion the required hairstyle will be the opposite of what you have. In each epoch only one kind of hair is permitted, and whatever you've got, that ain't it.

The obligation to follow the trends has unfortunate consequences, for instance when the fashion for bare midriffs comes face to face with the obesity epidemic. Murphy's Laws see to it that at the one time in human history when it would be a good idea to hide it, we have to flaunt it. The results are often tragic to behold, but there is nothing that proudly independent young people won't do to look like each other.

Sometimes the fashion meme can stabilize, perhaps for centuries. It is rechristened 'tradition', and occasionally causes some very strange visions to waft into view. In England the ultimate fount of all wisdom is a Law Lord, guardian of the nation's justice system. Sober and respectable, this august sage has to dress up in a way which, when you come to think of it, looks just a little like auntie Dora down the corner store, with hair up in curlers, and still in her dressing gown. But to change it would apparently threaten the very foundations of civilization.

When students graduate they are supposed to wear a flat piece of wood on their head all day. They could, if they wanted, decide *en masse* that this was a silly way to advertise their intelligence, but the meme is strong, and the mortar boards go on.

THE MURPHY'S LAWS OF CORPORATIONS

Memes permeate a business as 'corporate culture'. Every organization, from McDonalds burgers to McTavish's autospares, has its own culture, which reflects the nature of the man who runs it. If you want to get on with

McTavish you need to do things his way. If you want to
get on in McDonalds, presum-
ably the same applies. But who
is this McDonald chap that we
should look up
to? The corpo-
ration is so
vast, it must be
hard to work
out what is
the essence of
M c D o n a l d -
icity, and easy
to get it wrong.

Murphy's Laws will undoubtedly be found here.

Corporations move in mysterious ways. We can find a para-
ble in the world of slime mould. This fungus, *Dictyostelium*,
is found on rotting wood. It feels slimy because it is basi-
cally a mass of amoebas slithering over each other, munching at the bacteria in the wood, pausing to have a family
life, which it does by stopping for a moment, splitting down
the middle into two, then munching onwards. Each amoeba
is autonomous, its own master.

When the food runs short, however, something strange
happens. A chemical signal spreads through the colony and
compels them to turn towards the centre, where they form
a slug shape and start moving up the log. Each amoeba, if
you were to ask it, would stress that it is still as much an
individual as ever it was. But it is clear to an outsider that
it has a higher calling. It is part of the 'corporation'. Within
the slug body it has a specific job, depending on where it
finds itself. If it is at the bottom it has to keep the slug

flowing forward. If a little while later it finds itself at the front, its more sensitive side is called for – it needs to sniff out the way forward. When this slug has found the highest point of the log, another change comes over the colony. It grows a stalk. On the end of the stalk it grows a pod in which a small group of amoebae sit. While the stalk withers and dies, the pod is picked up on the next breeze and blown to a new location, where it starts the life cycle over again.

To BOLDLY GO...

Munch   Munch

So a seemingly independent amoeba can become part of the slime mould corporation and perform several coordinated and difficult manoeuvres, before expiring for the sake of the colony, having ensured the survival of a few at the top. Sound familiar?

In corporations and in governments there is a structure whose task is to make sure everything that is done preserves the purity of the corporate. It is called a **committee**.

Committee: a place where good ideas go to die

The task of a committee, according to fable, is to encourage good, new, imaginative ideas to pop into a back room, where they are quietly throttled. J.F. Kennedy called them 'twelve men doing the work of one'. Milton Berle said, '. . . a group that keeps minutes and loses hours'. Arthur Goldberg said, 'If Columbus had an advisory committee he would probably still be at the dock'. Why do committees get such a bad press?

A committee is part of the corporation's immune system. New ideas are always considered dangerous at first and treated with extreme caution, like possible viruses. The committee acts as a corporate lymph node, investigating the alien idea. If it seems to fit in with the corporate culture it is absorbed, if not it is deferred to a later meeting, and then again, until it withers and dies.

To say that all committees are deserts of the soul is to malign the many committees that know what they are there for and function extremely well. However, when a group has no clear aim, the result can be that a splendid idea is tugged in all directions until it is horribly mangled, then given a budget.

It is at the same time sweetly democratic and maddeningly frustrating when no one committee member can dominate. Everyone defers, is friendly, even-handed and supportive, and in the end they find an answer which can't possibly be the best, but could perhaps be the least-worst.

A committee is a group of people who can't decide anything on their own, but together can decide that nothing can be done

(Fred Allen, 1894–1956, American broadcaster)

There is also an unnerving feeling at the heart of the committee that everyone is listening out for a mystery voice. Everyone around the table knows in their bones that the corporation that spawned them wants it to reach a certain conclusion, which it cannot tell them, based on criteria which it will not reveal. So the committee is playing a game of Chinese Whispers with an invisible, inaudible extra player, the meme in the team. In many cases the committee's default belief is that management would really rather like it if nothing too dramatic happened. And so it happens. The one thing the committee cannot do is commit.

The approach to the problem is more important than the solution

Internal consistency is valued more highly than efficiency

If a problem causes many meetings,

the meetings eventually become more important than the problem

(all from Arthur Bloch, *Murphy's Law*, 1977)

There is no gene for committeeing. Committee procedure is a meme which gets handed down the generations. In place of DNA's basic building blocks of adenosine, guanine, thymine and cytosine (A, G, T, C), committees have agendas, quorums, proposals, amendments, any-other-businesses and motions to adjourn.

A conclusion is the place where you got tired of thinking

(Matz Law, quoted by Arthur Bloch in *Murphy's Law*, 1977)

Whether the committee is an eternal thing or set up for a specific purpose it must, on occasions, reach conclusions. Moreover, they must look like conclusions and not an indecisive fuzz. This is where a report, full of jargon, buzzwords, frequent use of optimistic phrases like 'going forward' and 'empowerment' will put up a good smokescreen, project the company culture forward, add brownie points to everyone's CV. In the end . . .

The means justify the means

SUCCESS AND MURPHY'S LAW

Can Murphy's Law help you work your way to the top? Jean Giraudoux offered some famous advice:

The secret of success is sincerity – once you can fake that, you've got it made

(Jean Giraudoux, 1882–1944, French playwright)

We want to be able to understand a person fully, deep down, and just about the only place where that can be done is by studying their face. We have reserved a large area of brain for that small square foot of flesh. How much can we psychoanalyse a person simply by looking at them? How easy is it to read a face? All humans share the same repertory of 7,000-odd facial gestures. One of these, the 'social smile', is used more than most, because, as Rita Carter points out in *Mapping the Mind*, it allows us to lie. The social smile is quite different from a natural smile. Brain scans show that different parts of the brain are involved in generating it, and different muscles are used to make it. The genuine smile uses tiny muscles around the eye which the social smile does not, and tends to linger for longer than the social smile. If you can learn to fake those muscle movements, you have indeed got it made.

It seems most of us are genuinely hopeless at telling the difference between a real and a false smile. In 1990 Paul Ekman and others at the University of California told two groups of nurses that they wanted to test how bright and cheerful they could be in all circumstances[5]. They then showed them a film which they were to describe to a panel of judges in a chirpy, cheerful way. The first group saw a pleasant film, one they could genuinely smile about; the second saw a film in which people were suffering horrendous injuries. They were all interviewed by experts in studying body language; psychologists, judges, detectives, secret service agents and customs officers. Very few of the experts could tell who was giving the false smile.

An ounce of image is worth a pound of performance

(Laurence J. Peter, 1919–1988, American educator and writer; *The Peter Principle*, 1969).

In 1946 Solomon Asch demonstrated elegantly what little difference there is between a good image and a bad image[6]. He gave people a list of adjectives that described a fictitious person. For one group the adjectives were: intelligent, skilful, industrious, warm, determined, practical and cautious. For the second group the adjectives were the same, except the word 'warm' was replaced by the word 'cold'. The difference in attribution was remarkable. The one person was judged to probably be generous, popular, sociable and humorous, the other was judged parsimonious, unpopular, solitary and dour. Whatever else you may be, it seems, it pays to be warm.

When we judge people we base our conclusions on some pretty ropey evidence. Psychologists refer to **cognitive miserliness** (the tendency to lump many different traits together), **naive science** (the tendency to jump to simplistic conclusions) and the **halo effect** (the tendency for one apparent trait to influence others, as the word 'warm' did, above). The outcome is that we end up forming a wide range of opinions based almost entirely on guesswork.

- **Cognitive miserliness**: Many people use footwear as a benchmark of character; good soles, good soul. That may be an extreme example, but most of us are convinced that fat people are jolly, high foreheads are a sign of intelligence, close-set eyes are a sign of untrustworthiness. Handsome faces conceal handsome personalities. And, as Dorothy Parker put it, 'Men seldom make passes at girls who wear glasses'.

- **Naive science**: Just your choice of name can get people to reach a simplistic conclusion; in an experiment reported by Adler[7] 464 psychiatrists were asked to give an assessment, based on a one-page description, of a patient who had apparently committed an assault. They all got the same description, but for half the patient's name was Matthew, while for the other half he was Wayne. Wayne got a consistently more damning assessment. In Shakespeare's *Romeo and Juliet*, Juliet asks, 'what's in a name?' The answer is; quite a lot really. That's why Shakespeare didn't write 'Wayne and Juliet'.

- **Halo effect**: Surround yourself with nice things and some of it will rub off on you. It really will.

Nothing succeeds like the appearance of success

(Christopher Lasch, 1932–1994, American social critic)

The selection of the right paraphernalia to aid the halo effect has never been more urgent. The complexity of culture creates a cornucopia of collectables to cling to. Ring-tones, nose-rings, bling-bling, roll-necks, Rolex, Rolls-Royce, hair-styles, Air Miles, big smiles . . . All these will express the real, successful you. So gather the trinkets about you like a bower bird.

A rich man's joke is always funny

(Thomas Brown, social commentator)

Being associated with the right people is important, too. Just being in the room for five minutes with an A-list celebrity adds a lustrous glow to your life. Your mirror neurons zing.

How much we link to the things and people around us was demonstrated in a very subtle experiment[8]. Students were asked to sort out a collection of words. Some were given neutral words, some were negative age-related words, such as 'decrepit' or 'arthritis'. When they had finished they were thanked and asked to go. They must have wondered, as they walked to the lift, what the whole thing was about. In fact they were being timed as they walked. The people who had been sorting age-related words walked to the lift significantly slower. Some of their task had rubbed off on them, enough for them to have 'aged' slightly.

If we're not going to have younger people feeling aged in our company, we'd better keep young ourselves. The multi-billion industry devoted to youth-enasia testifies to that. Sales of 'age-defying cream' are up, up, up.

It is even possible that our memes are being joined in this conspiracy by our genes. The controversial theory of neoteny points out that humans are becoming progressively less hairy, with larger eyes set further apart, just like a baby. Could it be that we're all evolving baby-like features, which make us look more attractive?

Here are a few general office rules from many sources. Obey them and you will have everybody around exercising their naive science, halo effects and cognitive miserliness on your behalf;

- Never walk around the company offices without a piece of paper in your hand.

- Never be the first to yawn.

- Never conduct negotiations before 10 a.m. or after 4 p.m.; before 10 a.m. you'll appear anxious; after 4 p.m. they'll think you're desperate.

- Always let the phone ring three times before answering it.

- Keep a look out for some good news to be the bearer of.

- Be out of the building when bad news arrives.

- Make sure you are seen with important people.

- Listen to jokes all the way through without blurting out the punchline.

- When told a piece of strictly confidential news, be assiduous in spreading it around.

- Find the boss charming and amusing.

- Address total strangers by their Christian name within 15 seconds of being introduced.

- Give out business cards to everyone, including family.

- Never speak in a lift.

- Never let the boss know you are cleverer than him.

- Never actually use the word 'no'.

- Or the word 'actually'.

- Have a massive hangover even though you never drank a drop.

- Every now and then try the boss's suggestion.

POWER, RANK AND MURPHY'S LAW

You want your views to prevail, you want to make your mark in your group, show your leadership qualities, get promoted, but there's a problem:

Anybody who's popular is bound to be disliked

(Lawrence Peter Berra, known as 'Yogi')

Berra here sets a puzzle for workers everywhere. To get on in the corporation you need to be popular, yet success requires cunning and aggression.

You find that warmth works (see p. 134). So be warm. Be damn warm! Everyone else will be trying to be warmer. Eventually one of you will be warmer and friendlier than all the rest, and they will be the most popular, and then it's a matter of time before the others gang up in a warm and cuddly way, and spike them.

Nobody is entirely sad at the failure of a friend

(Gore Vidal, b. 1925, American novelist, playwright and essayist)

With friends like this, who needs the Ebola virus? Gore Vidal's comment may be a tad cynical, but he's right on the money with this one. It's a paradox, that when a friend or workmate flops, half of you is saddened and half is a little gleeful.

There are two parts of you working in opposition, it seems; one part is the empathizing, team-building, warm you, with mirror neurons reflecting the trials of your buddy. The other part is the ambitious you, in a constant struggle to be top of the pecking order.

Any man more right than his neighbour constitutes a majority of one

(Henry David Thoreau, 1817–1862, American author, poet and philosopher; *Civil Disobedience*)

Here we enter the testosterone zone. Testosterone[9] has a lot to answer for. This hormone has been fingered as the main source of aggressive behaviour in males. From the moment testosterone production starts in men, at the age of about 13, rivalry between them begins: fights, arguments, sulks, depressions and boastings all buffet the family or workplace as the young bucks throw their weight about.

Why does adolescence happen? Wouldn't the world be so much more peaceful without it? Is it purely to make life difficult for everyone? Well, no, in the end it's to make life simpler.

A problem that grew alongside the evolution of mirror neurons was dealing with all the neural noise that came with it. If everyone's mind reflected everyone else's equally, then everyone's idea would be as good as everyone else's. Each brain would be flooded with waves of empathy from all around, as everyone related to everyone else. A raucous racket of rapport would result . . . too many generals . . . the broth spoiled by too many cooks . . . Many proverbs illustrate the problem.

The solution that evolution provided was the pecking order, with one acting as chief receiver and transmitter, and the others ranked below. This is a far more efficient system, and accounts for the brilliantly directed nature of teamwork; a team focuses on a task 'single-mindedly'. Everyone has their role within the group and knows their place.

Testosterone helps to sort out the ranking. Before a competition or fight between rivals, testosterone levels in their bloodstreams rise. After the result the winner's testosterone level rises further, the loser's falls. In monkey troops, testosterone levels exactly reflect the pecking order: High testosterone levels at the top, lower ones at the bottom. In unstable groups, where there is a lot of fighting for status, there is a big difference in testosterone levels between top and bottom ranks, but in more stable groups testosterone is no longer needed to mediate the ranking process, and levels are even throughout.

Nice guys finish last

(Misquote from Leo 'the Lip' Durocher, baseball manager, 1946)

Apes and chimps, our nearest relatives, have a well-documented social structure. The pecking order among males is ritualized, with aggressive behaviour only breaking out when a change in the order is in debate. These tussles can get quite rough, but are seldom life-threatening, and usually take the form of ritual displays. (Truly aggressive behaviour is reserved for when they hunt or fight the neighbours.)

The gestures used by office workers when they are under pressure are inherited from chimps, but now reduced to symbolic status; a raised finger replaces the waving club, an oversized smile replaces biting, tossing the biro on to the desk replaces uprooting a tree and flinging it across the forest, sarcasm replaces the spear.

And, yes, it can also get violent. A study conducted by the US Workplace Violence Research Institute in 1995 found that, 'every workday, an estimated 16,400 threats are made, 723 workers are attacked and 43,800 are harassed.' A 1996 Gallup telephone survey, also conducted in the USA, found that of the 1,000 adults they interviewed, 25 per cent reported being 'generally at least somewhat angry at work' (*Globe and Mail*, 23 August 1999).

In chimp troops the top of the pecking order is the dominant male. The other males defer to him, and he has the preferment of the females. In the finance houses of Wall Street his equivalent is affectionately known as the 'Big Swinging Dick'. (The big chief dominant male in the USA is the president. The deaf sign for 'president' is a thumb protruding through two fingers – a very big swinging dick, if scaled up.)

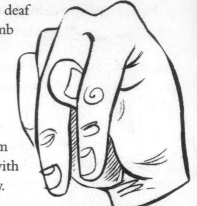

Below the BSD the pecking order runs all the way down. The one at the bottom has to ingratiate himself with everyone – he's the nice guy.

Last guys don't finish nice

(Stanley Kelley, Princeton professor, quoted 1976)

When *Homo habilis* had a bit of a tiff, the loser could take it out on a tree, a rock or a passing goat. Life in an office block is more constrained. Take it out on the office furniture and you get handed a repair bill, so our loser has to button it up, which is a problem.

Mr Nice is not feeling nice today. He has just come away from a 'blamestorming' session with the head of department. It was a 'bruising experience' but he is not bruised. The boss 'wiped the floor with him', but he remained seated throughout. The 'fur was flying', but no fur actually flew. In fact boss smiled throughout. Perhaps a little too much.

Although the gestures were muted, Mr Nice's fear centre, his amygdala, was in full flow. It recognized the signals from tens of thousands of years back. Boss was not a happy bunny, but this being the twenty-first century, he

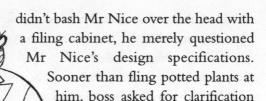

didn't bash Mr Nice over the head with a filing cabinet, he merely questioned Mr Nice's design specifications. Sooner than fling potted plants at him, boss asked for clarification about the budget overspend. Rather than jumping on the desk, roaring and beating his chest as Mr Nice left, boss just mentioned that wearing brown shoes with black trousers wasn't quite '*comme il faut*'.

But the effect was the same, and Mr Nice's reactions, unchanged since Palaeolithic times, were the same: his amygdala caused the endocrine system to secrete stress hormones such as adrenaline, which caused the blood to flow away from the skin and towards the heart and muscles, stopped digestion, increased heart rate, focused the attention mechanisms, and generally prepared him to dodge flying filing cabinets. So while Mr Nice sat listening to boss he was visibly pale, with a pounding heart, eyes strained open and pupils dilated. At the same time he behaved in ways which even a rat would recognize, head bowed, movements small and slow, speaking little, making eye contact not at all.

If this happens too often, the effect of the hormones on Mr Nice's body will be long-lasting. If he continues to find himself in stressful situations his food will remain undigested

– his arteries will clog with fat, he will suffer from high blood pressure and the risk of a heart attack, and high sugar levels could trigger diabetes. He could develop anxiety-related allergies, lose sleep, show erratic and aggressive behaviour towards his family at home. His brain will physically change shape; his hippocampus will shrivel up. He will die young. So either Mr Nice turns nasty, or he has a nasty end.

Or he leaves the company, which would be a pity, because Mr Nice is as vital to the success of that team as the BSD. Someone has to be at the bottom of the heap. Boss should realize this and look after Mr Nice. If he goes, then the battle for his successor at the bottom will begin. With every change of personnel the rivalries are reawakened and the fighting starts up again. It's all very bad for everyone's endocrine system.

Not all businesses are battlefields. We shouldn't forget that in most organizations the pecking order is stable, testosterone levels are low, the atmosphere is settled. Largely this is due to another Mr Nice accepting his role as a small fish in a pleasant enough pond.

Every Beatles needs its Ringo

The Ringo Law is named after Ringo Starr, the nice-guy Beatle. Ringo was a lovely lad, but he wasn't all that useful, and he didn't look

pretty. And he couldn't sing. And he wasn't even that good on the drums. A lot of Beatles fans wondered why they didn't get rid of him and find a really good drummer, of which there were plenty. So what if they had replaced Ringo with a really flashy fella, who wrote and sang, danced like Nureyev and had a PhD in economics? He would have fought for a higher place in the pecking order and the Beatles would have ripped themselves apart within a year. Ringo was the essential nice guy, happily finishing fourth of the Fab Four.

Here's a farmyard tale that shows it up. Some friends had four hens, with a clearly defined pecking order, Fat Betty at the top, Gurgle at the bottom. But they only laid three eggs daily between them, and Gurgle was the backslider. She was replaced by a fine, feisty new hen, known as a good layer. Feisty Hen immediately set about fighting her way up the pecking order. The result: two eggs daily.

MANAGEMENT AND MURPHY'S LAW

The modern middle manager has the task of making sure his or her department is happy with itself and in harmony with the company – if you like, they must define what jigsaw everyone is doing. But it is a thankless task, to keep his or her department steady in a sea of change. Back in the jungle, tribes of chimps stay pretty stable. Everybody knows everybody else, and has done since they were born. They are kith and kin. The pecking order in the family is sorted. Not so in the modern industrial world. Corporations expand and contract, and as they do new tribes, departments, pop into existence, establish their pecking orders, then pop off again. Brand new companies emerge from the mists, merge and demerge. New conglomerates conglomerate, then deglomerate. With each new office or factory new teams have to be hired from scratch. Total strangers must establish their pecking orders, which are not necessarily the ones imposed from above.

In such an ambiguous environment, Murphy's Laws proliferate.

In a hierarchy, every employee tends to rise to his level of incompetence

(Laurence J. Peter, 1919–1988, American educator and writer;
The Peter Principle, 1969)

The Peter Principle is beautifully easy to understand: a manager does a good job, so he or she gets promoted up a level. Up and up they go, until they get promoted to a job they can't do. Senior manager notices they're not doing so well, and doesn't promote them any further. This ensures that in the end every responsible job is held by someone who can't do it.

When in doubt, mumble; when in trouble, delegate; when in charge, ponder

(James Boren, American government bureaucrat
and founder of the Apathy Party)

This is the motto of middle management. Constantly caught in the crossfire, their only wish is to survive. Senior management want them to be decisive and single-minded, yet moderate and flexible; sticklers for discipline, yet relaxed and easygoing; to cut costs and speed production, yet keep quality as high as ever. The people below the middle manager want simply someone who knows the craft. Research in South Africa into what qualities were most admired in middle management[10] showed that the line managers respected technical know-how, while the directors favoured people skills. You're damned whatever you do. To fudge and mumble is about the only way out.

A life raft for the drifting manager is the corporate culture.

> Anyone can make a decision given enough facts; a good manager can make a decision without enough facts; a perfect manager can make a decision in complete ignorance

(Spencer's Law of Data, from Arthur Bloch's *Murphy's Law*, 1977)

The perfect manager is able to make a decision in complete ignorance because he understands the company's culture instinctively. Facts will only get in the way. He does everything according to the book. His actions will support the status quo long after the quo has lost its status. His decisions will be elegant, symmetrical, unblemished by ugly reality. He will win the praise of his masters and the adoration of his underlings. He will drive the company into the buffers.

J.K. Galbraith, in *The Age of Uncertainty*, described the decline of the British army in the years before the First World War in similar terms. The terrible blunders of that great folly were not due to the soldiers, who were 'lions led by donkeys', but caused by decades of decay among their officers. In times of relative peace, at the end of the nineteenth century, the way to promotion throughout the British Empire was not through military ability, but by expressing the right degree of national pride. Capable generals who were critical of British foreign policy would not get on, whereas those who praised the Empire culture reached the top however bad they were at their job. In addition, an incompetent general would be sure to promote junior officers who wouldn't show him up; in other words, he preferred underlings who were more incompetent than he was. These in their turn would make similarly appointments below them. Thus incompetence rose up through the ranks to the highest echelons. Their ineptitude remained unnoticed until they suddenly had to make serious military decisions, as the war with Germany got under way. By then it was too late for the tommies in the trenches.

In every organization there will be one person who knows what is going on. This person must be sacked

(Conway's Law, quoted by Arthur Bloch in *Murphy's Law*, 1977).

MURPHY'S LAW AND TERRITORY

Early man had well-defined territories. No rival tribe came near without expecting skirmishes on the boundary. Nowadays, as post-Palaeolithic man roams the concrete jungle, he has many confusions about which tribe he's in, and where the boundaries have gone. Humans have not evolved fast enough to deal with this brave new world.

An example given in *The Meaning of Liff*, by Douglas Adams and John Lloyd, describes encounters in corridors. Corridors are a new invention. Passing someone involves approaching them directly, which Palaeolithic man sees as threatening. Getting past them means invading their personal space, which Palaeolithic man sees as provocative. And are they friend or stranger? Palaeolithic man would never let them on to his territory without a fight. Adams and Lloyd outline a strange etiquette which has evolved in the company corridors:

- **Corriearklet** (n.): The moment at which two people, approaching from opposite ends of a long passageway, recognize each other and immediately pretend they haven't. This is to avoid the ghastly embarrassment of having to continue recognizing each other the whole length of the corridor.

- **Corriedoo** (n.): The crucial moment of false recognition in a long passageway encounter. Though both people are perfectly well aware that the other is approaching, they must eventually pretend sudden recognition. They now look up with a glassy smile, as if having spotted each other for the first time, (and are particularly delighted to have done so) shouting out 'Haaaaaalllllloooo!' as if to say 'Good grief!! You!! Here!! Of all people! Well I never. Coo. Stop me vitals, etc.'

- **Corrievorrie** (n.): Corridor etiquette demands that once a corriedoo has been declared, corrievorrie must be employed. Both protagonists must now embellish their approach with an embarrassing combination of waving, grinning, making idiot faces, doing pirate impressions, and waggling the head from side to side while holding the other person's eyes as the smile drips off their face, until with great relief, they pass each other.

- **Corriecravie** (n.): To avert the horrors of corrievorrie, corriecravie is usually employed. This is the cowardly but highly skilled process by which both protagonists continue to approach while keeping up the pretence that they haven't noticed each other – by staring furiously at their feet, grimacing into a notebook, or studying the walls closely as if in a mood of deep irritation.

- **Corriemoillie** (n.): The dreadful sinking sensation in a long passageway encounter when both protagonists immediately realize they have plumped for the corriedoo much too early as they are still a good 30 yards apart. They were embarrassed by the pretence of corriecravie and decided to make use of the corriedoo because they felt silly. This was a mistake as corrievorrie will make them seem far sillier.

- **Corriemuchloch** (n.): Word describing the kind of person who can make a complete mess of a simple job like walking down a corridor.

Many court cases have been brought by furious neighbours willing to bankrupt themselves with lawyers' fees to contest tiny changes to boundary fences. In 2003 George Wilson was gunned down in the doorway of his Lincoln home by his next-door neighbour. They had been arguing over a privet hedge that ran between their front gardens. The territorial instinct comes down intact from our chimp ancestry. As far as this aspect of evolution is concerned, we aren't out of the woods yet.

Her pile is mess; your pile is work in progress

Even the home is full of fuzzy boundaries. They aren't simple;
houses aren't often divided in the way they do in cartoons,
with a line painted down the middle, so he gets to use the
right hand side of the stove and table and sofa, watches
the right hand side of the TV, etc. and she gets what's left,
in a very real sense. But there is an invisible line, a virtual
boundary, around what he considers his fiefdom, and around
what she won't tolerate being tidied. Her kitchen, His study,
Her flowerbeds, His rockery, Her runaround, His hatchback.
He will look at her 'outpost' of bits and bobs on the table
with complete incomprehension; she will be equally mysti-
fied by his pile of nonsense. Each can see the importance
of their own pile, and will dismiss their partner's.

The invisible boundaries of everyday life.

Companies have equally fuzzy boundaries, both within and without. Their reaching out to claim territory beyond the normal boundary is legendary. Companies stretch patent and copyright laws to the limit to expand the companies' territories and guard against invasion.

- Harrods, the famous London department store, sent a team of lawyers to the other side of the world to sue a certain Mr Harrod in New Zealand, who had opened a general store called, unsurprisingly, Harrod's.

- Mike Rowe set up a computer software company called, for a brief while, Mike Rowe Soft, until the writs began to arrive.

- If you wave your arm in a certain way you could fall foul of Alan Shearer, who has copyrighted his characteristic gesture of goal-scoring victory.

- In Hollywood the building used by the Walt Disney Studios for film production is half the size of the building devoted to copyright protection. Every mouse character which appears on the animation scene is sniffed over to see whether it should be classed a 'pest'.

Here is a case where two companies laid claim to a piece of fruit. Steve Jobs had set up the Apple Computer Inc. in California in 1975 and his company was spreading eastwards fast. Meanwhile in England the Beatles were investing their earnings in a music company, called Apple Corps (because John Lennon liked the pun). As each company looked at expanding across the Atlantic, they came to realize that their names could clash. People might buy a computer expecting it to play rock and roll music, they thought, or pick up a record in hope of doing the accounts with it. Several committee meetings later they came to an agreement: Steve Jobs promised not to sell music if the Beatles promised not to branch into computers. The two protagonists were going to be able to coexist, like good hippies should.

In fact, Steve Jobs discovered soon after that he wanted to use sounds as alert signals, and one of the alerts was a chord. Was this music in any legal sense? Several committees pondered. Steve cut through and issued the sound. It's still there today, called 'sosumi' ('so, sue me!'). The Beatles didn't sue.

The appearance of the ipod and itunes changed this bonhomie. With Apple now seriously involved in the music business the two beasts got right down to slogging it out.

LANGUAGE BARRIERS

The use of a secret language understood only by a small circle of friends creates a kind of sound territory. Only those within the verbal boundary are members of the tribe. Newton published his *Principia* in Latin deliberately to stop most people having the faintest chance of criticizing it. The famous eleventh-century Arabian scientist, Geber, wrote in a secret code for fear of persecution. Our word 'gibberish' comes from that. Modern-day doctors follow the same principle, using Latin to make simple conditions sound complex. It is somehow comforting to tell the doctor that you've got a bad back, and have him diagnose lumbago ('lumbago', incidentally, is Latin for 'bad back'.)

Corporations use a sort of pidgin in the same way. An opacity of language inspires humility in the listener, who thinks he has much to learn. As Leo Rosten and John Peers point out in *1001 Logistical Laws*,

Most people confuse complexity with profundity

(Leo Rosten, b.1908, 'renaissance mensch')

Here are some words gleaned from the many thousands on display at Buzzwhack, the website dedicated to tracking all the new buzzwords, all of them tantalizingly close to being understandable:

disambiguate, functionality, agreeance, back-sourcing, conversate, granularity, infotisements, insourcing, partnering, intrapreneurial, strategery, togethering, operationalize, planful, deliverables, leveraging, de-risk, deconflicted, defenestration, efforting, exceedance, face time, going forward.

The Freeman Institute (http://www.freemaninstitute. com/bingo.htm) has developed Lingo Bingo to keep you alert during pep talks. The winner is the first to complete a row or column.

Synergy	Strategic Fit	Core Competencies	Best Practice	Revisit,
Bottom Line	Result-Driven	Out of the Loop	Benchmark	Knowledge Base
Ball Park	Game Plan	Client Focus[ed]	FastTrack	At the End of the Day
Win-Win	Value-Added	Empower[ment]	Leveraging	Take That Offline
Paradigm	Proactive	Mindset	24/7	Think Outside the Box

Lingo Bingo: the first to complete a row or column is the winner.

MURPHY'S LAWS AND OBEDIENCE

For every heap there is only one top. Those at the top will tell those below what their jigsaws look like. And those below will agree, whatever they were thinking a minute ago.

The meek may inherit the Earth, but not its mineral rights

(Zillionnaire J. Paul Getty, 1892–1976, America's greatest
oil magnate – he would say that, wouldn't he)

If the meek are going to inherit anything they must be ready to fight in defence of their meekness. But fight is what they will not do. They know their place. They will do what they're told.

J. Paul Getty

Stanley Milgram gave a startling demonstration in obedience in 1963, in an experiment which stretched the ethics of psychological research to the limit[11]. Pairs of subjects were asked to help with a task, an investigation of the effects of punishment on learning. One was chosen randomly to be the teacher and the other the learner. The learner was taken into an adjoining room and strapped into a device which delivered an electric shock to his arm. The teacher watched him through a glass window. In front of him was the button with which he was going to administer the shock and a dial with which to increase the voltage. The dial was marked from 15 to 450 volts, and verbal descriptions next to the dial ran from 'slight shock' through 'intense shock' to 'danger: severe shock'. The last space of all simply had 'XXX'. The learner had to sort cards. If he made a mistake, the experimenter instructed, the teacher had to give him a shock, and each shock had to be 15 volts greater than the last.

And so the experiment began. The teacher administered shock after shock, turning the dial up each time, to an increasingly agonized learner. What the teacher didn't know was that the 'learner' was actually a stooge. It was all a fake. He was receiving no shocks at all, but reacting to a signal light hidden in a corner. The teacher, unaware of this, kept raising the voltage, until the learner seemed to be knocked unconscious by the severity of it. When the experimenter told the teacher to increase the voltage and shock again. He obeyed. All of the teachers continued to shock up to 315 volts, by which time the learner was screaming for him to stop. Two thirds of them went right up to the last mark, 'XXX'. In nearly all cases the teachers showed great anguish, attacked the experimenter verbally, trembled and sweated. But still they obeyed. The authority conferred by the white lab coat has never been revealed to such a degree.

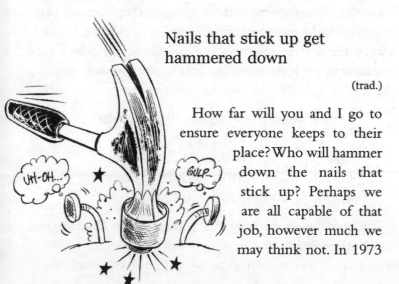

Nails that stick up get hammered down

(trad.)

How far will you and I go to ensure everyone keeps to their place? Who will hammer down the nails that stick up? Perhaps we are all capable of that job, however much we may think not. In 1973

Zimbardo and other psychologists set up a prison simulation to find out[12]. Twenty-four volunteers, drawn from all over America and selected for their emotional stability, reasonable health and average personality, were divided randomly into 'prisoners' and 'guards' in a two-week study of prison life. The prisoners were prepared in the same way that real prisoners are, strip-searched, deloused and issued with prison clothes. (They were excused having their heads shaved, but wore nylon stockings over their hair.)

The twelve prisoners were taken to the prison cells and waiting guards, and almost immediately the abuse began. The guards, hesitantly at first, but with increasing enthusiasm, began to systematically humiliate the prisoners. The punishments quickly became so frequent and imaginative that the experiment, though planned to go on for a fortnight, was abandoned in six days. Afterwards many of the guards described the pleasure they got from exercising power over others. Often they were surprised at their own cruelty. Even though at the beginning they were all play-acting for the experimenters, adopting roles based on what they thought prison was like, a point came when they started to get real, where the Palaeolithic showed through.

> When someone says something foolish,
> they're foolish. When 5 people say
> something foolish they're foolish
> 5 times. When 5,000 people say
> something foolish, you're foolish

It is hard, being the only one in step, as Mrs Thatcher said for many years.

In several experiments during the 1950s, Solomon Asch found out how hard[13]. Each volunteer was told he was one of 11 volunteers who were going to make a series of judgements about which of 3 comparison lines was the same length as a standard line. He was put in a room with 10 other people who he thought were other participants, but were actually stooges. The subject was sat at

Asch's Experiment: eleven subjects had to select the matching line, but ten of them were stooges.

the end of the line of judges, so all the others gave their vote before him. To begin with they all agreed, but on a secret signal each of the 10 stooges in turn opted for the wrong line. When it came to number 11, he hesitated, looked again, and voted with the others. The rigged vote happened 12 times in the test. Out of 50 volunteers tested, three quarters voted at least once against his better judgement. One of them did it 11 times out of 12. You can imagine he would get on well in the corporation of his choice.

In a hierarchical system, the rate of pay varies inversely with the unpleasantness of the task

(First Law of Socio-economics, Arthur Bloch, *Murphy's Law*, 1977)

A stunningly Murphoid discovery, with a very good scientific provenance. Festinger's 1957 theory of **cognitive dissonance** explains that if a person finds themselves holding two contradictory thoughts they will do anything to reduce the 'dissonance' created. This includes, if necessary, changing their belief system a little. In a famous experiment[14] college students were given extremely dull and repetitive jobs to do, then given either $20 or $1 as payment.

Who reported greater job satisfaction? You would think that those paid the most would be happiest, but the results were quite the reverse; those paid $1 expressed greater job satisfaction than those paid $20. This is counter-intuitive. Cognitive dissonance, though, digs out the truth. The poor students who found they had lost an afternoon and gained only $1 would not allow themselves to believe they had been duped. Unconsciously they boosted the value of the work they'd done so they could feel good about themselves.

If we are all so compliant, what does that mean for the complex business of choosing our nation's leaders? How does democracy work? Churchill said that democracy was the 'worst form of government, apart from all the others'. It's all really very straightforward:

Democracy consists of choosing your
dictators, after they've told you what
you think it is you want to hear

(Alan Coren, b. 1939, British writer and satirist)

Alan Coren tries to describe here the forces that lead
voters to vote the way they do. The psychology of voters
turns out to be very simple. Frighteningly simple, according to Josef Sznajd and Katarzina Sznajd-Weron, at the
Polish Academy of Sciences in Warsaw; people vote like
their neighbour. The Sznajd model of voting is based on
the idea that, in certain aspects at least, human society will
behave pretty much like a bar magnet[15].

Each iron atom in a magnet is itself like a minute
magnet, with a north and south pole. When iron is heated
the atoms jiggle, spin and roll in all directions. As the
iron cools they will tend to settle the easiest way, which
is with north and south poles in alignment. It is not
universal; whole clumps will end up in one direction,
others in another, with an area of chaos on the border.
The way any one atom turns will be heavily influenced
by what its neighbours do. When this model is set up
against voting patterns, the match is uncanny. People will
tend to vote the way their neighbours do. It's as simple
as that. (The areas of chaos in the magnet become 'swing
states' or 'marginal constituencies' in the election.) As
anyone knows who has tried to hold to a political view-
point when surrounded by opposite opinions, this is
reality.

Of course there are dissenting voices. As George Burns
said:

It's a shame that all the people who know
how to run the country are too busy driving
cabs and cutting hair

(George Burns, 1896–1996, American comedian)

Those who know how mad it all is can take up occu-
pations where they can go mad in their own style, like
hairdressing or taxi driving. These little magnets will
always point the wrong way. Bless.

John Lennon's comment resonates for those who try to
point a slightly different way occasionally;

Life is what happens while you're making
other plans

(John Lennon, 1940–1980, Beatle)

CIVILIZATION AND MURPHY'S LAW

We can wonder how much monkeys, with their mirror
neurons, can learn from each other, and why only one
branch of the family went on to become us, with our
very special skills. But there is no doubt that language has
played a crucial part. The invention of language is one of
the major breakthroughs of *Homo sapiens*. Perhaps **the**
major breakthrough, because this meme can be used to
transmit other memes. It is the meme of memes.

A monkey who wants to teach another monkey how
to get sand off potatoes can't talk about it. She needs
sand, potatoes and water. She can't handle a hypothetical
potato and imaginary sand. We humans can use language
to discuss the problem all together in abstract and work

out the solution by phone link to the other side of the planet. We can look up in a book how the problem was solved 3,000 years ago. When we've finished sorting out the potatoes we can use our language skills for abstract debate about politics, philosophy, the meaning of existence, and round off the evening solving Fermat's last theorem. The monkeys have arguably got more done in the day, but we have had more fun.

We use the language meme to share our culture through the media, such as newspapers. Therefore look for Murphy's name on the by-line.

Everything you read in a newspaper is true, apart from the one story of which you happen to have first-hand knowledge

(Erwin Knoll, 1931–1994, American editor of *The Progressive* magazine)

Erwin Knoll's ironic surprise is mirrored by A.J. Liebling's response that:

People everywhere confuse what they read in the newspapers with news

(A.J. Liebling, 1904–1963, American journalist; *New Yorker*, 1956)

It would be far better to think of the newspapers as transmitters of memes rather than spreaders of truths. They listen to what you wish to be told, then tell you it. This guarantees sales. Any newspaper which doesn't more or less conform to the nation's culture is risking a serious fall in circulation, as the *Daily Mirror* found when its vocal opposition to the invasion of Iraq in March 2003 led to an instant collapse in sales. It quickly reversed its moral principles to claw back some of its readers.

A simple lie is better than a complicated truth

(Saki, 1870–1916 – pseudonym of Hector Hugh Munro, British author)

Orthodoxy is splendidly simple. Reality is so rough. Orthodoxy prefers the big statements, sweeping aside the small inconsistencies to sail majestically towards the grand fallacy.

Men occasionally stumble over the truth, but most of them pick themselves up and hurry off as if nothing had happened

(Winston Churchill, 1874–1965, British prime minister)

Amid the bluster and rhetoric, it is no longer certain where truth has gone. Perhaps we can find it among proper, serious scientists, who wear reassuring lab coats and deal with pure, empirical, testable ideas. The problem is that many scientists have a problem about simple truths; they don't believe there are any. They know that we've been wrong so many times, from ancient orthodoxies that put the Earth in the centre of the Universe to the recent flurry over cold fusion. So scientists tend to hedge their bets to the point of fetishism. For instance, while radical political groups opposed to genetically modified foods will proclaim, 'Frankenfoods will cause deformed babies and the end of civilization', a militant scientist will sound more like, 'the current state of evidence from peer-reviewed studies would tend to support the hypothesis that, on balance . . .' But who's listening to them?

One thing is for sure, the diffident scientist won't be writing the history books. That task is too important to be left to seekers after truth.

History is the version of past events that people have decided to agree on

(Napoleon Bonaparte, 1769–1821, French emperor)

Napoleon's observation is perhaps a little modest. The 'people' he refers to were himself and his chums. Hardly surprising that the 'people' decided to agree he was the saviour of humanity, and depicted him in the mass media of his day (paintings and engravings) as a hero equal to the greatest of Roman emperors. Clad in imperial purple, crowned in the victors' laurels, riding a fantastical horse. Essentially it was his-story.

History is largely the propaganda of the winning side. Look at the ordure heaped on Richard the Third, the misshapen hunchback played with such glee by

Shakespearian actors for over 400 years. Richard III didn't have a hunchback. It was the brilliant invention of Thomas More, an ambitious young man, given the job of writing a history book for Henry VIII. Henry's father, Henry VII, had defeated and killed Richard and Henry VIII wanted to legitimize his claim to the throne. When More saw a picture of Richard III which could conceivably be interpreted to suggest that Richard sat badly, he knew just what to do. His tale depicted Richard as a man warped in body and mind. And that is what has come down to us as the accepted truth.

Until the late twentieth century British history books portrayed the bringing of 'civilization' to India, Southern Africa and the Far East, always stressing that the armies of conquest were purely a philanthropic gesture, and the Raj was a giant charity. But of course everybody does it. Japanese history books were revised in 2001 to downplay their invasions of Korea and China in the twentieth century. Right now, Hollywood is busily rewriting the whole planet's history. Everywhere from Troy to the Titanic looks a little like California these days.

The story of history is the one which everyone accepts. But, as Murphy points out,

That which is accepted by everyone is bound to be false

(Paul Valery, 1871–1945, French critic and poet; *Tel Quel*, 1943)

Any widely accepted belief system is going to be agreeably easy to digest. But truth is never that simple. Understandably, many people of a slightly more subtle turn of mind find themselves at variance with the whole basis of the surrounding culture; religious in a secular state, Catholic in a Protestant state, a Shia among Sunni. It would be grand if it didn't matter. The national culture could include the whole range of spiritual beliefs, from those who reckon we're an accidental collation of molecules clinging to a piece of damp rock floating through an infinite amount of nothing at all, to those who believe the Universe was made as a present for us by a big chap on a cloud 6,000 years ago, and all stations in between. In spite of what we see in the media, most nations do actually toler-

ate a wide range of beliefs, but still in far too many nations only one belief system is tolerated.

I love Humanity – it's people I can't stand

Linus, in the Peanuts cartoon, sums it up for all of us some of the time, and for some of us all of the time.

The tale of our 'humanity' has to be a neat story, rounded off at the edges to make it all nice and tidy. History's kings and queens were good or bad, never nice in parts. The wisest person in the land is elected president, heads of corporations are honourable and generous, policemen honest, judges impartial.

Just don't look too closely.

Those who respect the Law and like sausages should not watch either being made

(Otto Von Bismark,1815–98, Prusso-German statesman)

SUMMARY

This chapter started with a simple structure of the brain: mirror neurons. Everything stems from there. Mirror neurons create memes, which generate our culture, and we rely on our cultural norms to inform us how we should think. One way or the other, it seems our tribe, clan, family, friends, church or state have us surrounded.

Perhaps you would like to escape? Murphy's Law has you boxed in:

You can't win

You can't break even

You can't even quit the game

(Allen Ginsberg, 1926–1997, American poet)

Freeman's Commentary on Ginsberg's theorem

Every major philosophy that attempts to make life seem meaningful is based on the negation of one part of Ginsberg's theorem. To wit:

- Capitalism is based on the assumption that you can win.

- Socialism is based on the assumption that you can break even.

- Mysticism is based on the assumption that you can quit the game.

Or perhaps not. Brain scans give a hint that there may be a way out.

The sense of inner contentment, which Buddhist monks claim to experience and which any operative sitting at their work-pod reckons is a myth bigger than the tooth fairy, turns out to be genuine. To find out, Andrew Newberg, a radiologist at the University of Pennsylvania, took a brain scan of a meditating monk, using single photon emission computed tomography (SPECT)[16]. On the one hand, this was a monumental undertaking. What could be less conducive to contemplation than being fed into the middle of a photon detection array and told to pull a string, which injects a radioactive tracer into the brain, at the moment of reaching a transcendental high? On the other hand, if anyone can, a Buddhist monk can.

And they did. The results clearly show a difference in the operating of the brain during meditation. The left prefrontal lobe, associated with concentration and positive emotions, lights up, and the parietal cortex, linked to orientation and positioning, showed reduced activity. So next time you are faced with the calm smile of a genuinely centred, happy individual, resist the temptation to wrestle them to the ground and help them access their inner misery, they really are like that.

He who laughs, lasts

7

The World Turned Upside Down

So, to summarize the book so far, is this what happens when you deal with a situation? . . .

1 First you look to see exactly what shapes and sounds, colours and odours are there in front.

2 Then you reach for your memories to identify what the shapes and sounds must be.

3 Then you connect each event or thing to all the others to make a coherent story.

4 Then you form an emotion which is appropriate to the scene; you grow a sense of fear, love or hatred towards what you see.

5 Then you fit everything into a social context, decide what society would consider an appropriate response and behave accordingly.

Unfortunately there are too many times when exactly the reverse happens. The tragic story below would never have happened if the psychological train of events had been the right way round.

On the evening of 2 September 1999, Harry Stanley, a 46 year-old man from Hackney, was shot dead by armed police in east London. He was walking home from the local pub, carrying a coffee table leg in a plastic bag, when officers from Scotland Yard's specialist SO19 firearms team shot him dead. It was because of the table leg that he died. And the reason for this can only be understood if we look at the processes above in reverse.

1 First, the **social context** was set in place. At the same time as Harry Stanley was leaving the pub, the police received an anonymous call claiming that an Irishman carrying a gun was setting out in the same direction. The hoax call caused a picture to be painted in the minds of the officers. This was Hackney, after all, a dangerous place late at night. The Irish had a reputation among the police for being villains (a relic from the bitter days of the Northern Ireland troubles).

2 The officers could easily form a range of **emotions** to colour that image: fear; certainly fear felt on behalf of the innocent citizens who they believed were under threat, but most particularly a fear for their own life. Also a feeling of honour and duty to put oneself in the firing line in order to protect society. A brooding suspicion of all things Irish may have added extra spice to the mix. (Harry was actually Scottish.)

3 And so the **connections** were made. Before the officers found Harry, a coherent story had been written: an Irish terrorist had been drinking all night in a pub, and had now set out to commit a murder with a gun, which he had hidden in a plastic bag. He had to be stopped. (The actual story couldn't be more different. Harry, a father of three, was recovering from a recent stomach operation. He was going to use the leg to mend a coffee table the next day. He wasn't drunk, he wasn't Irish, he wasn't interested in politics, he certainly wasn't a murderer.)

4 How did the officers' **memories** serve them when they caught up with Harry? They saw him carrying a long thin object concealed in a bag. Their memories from weapons training told them that the shape was about right to be a sawn-off shotgun. It all fitted into place.

5 Harry heard someone shout at him. He turned towards the sound, taking hold of the bag with the table leg in it. What did the police officer see? At the inquest he said he thought the 'Irish terrorist' was turning to shoot at him. Harry's movements after his operation would have been slow and stooped, which would have been further proof to the prepared mind of the officer. But nothing he could have done would have prevented him from being shot. The story had already been written.

Afterwards an officer looked in the bag. 'I had to stare' he said at the inquest. Of course. He was trying hard to see what his mind had prepared him to see – a gun, not a table leg. If his eyes continued to show him the wrong picture the consequence was going to be dire. He probably blinked a lot, and rubbed his eyes.

In the section that follows, on science, the Murphy's Laws all have perfectly rational explanations. That won't stop them from proving again and again the Law of Laws:

Whatever can go wrong, will go wrong

8

Pure Science

Many years ago I wanted to write a book which would explain all of Murphy's Laws, the pure, empirical science behind everyday glitches. After years of research I have discovered that to explain Murphy's Laws takes a pinch of hard science and a sackful of psychology. Ironically, just about the only Murphy which has a rigidly physics-based explanation is the one which is quintessentially Murphy – the toast always lands butter-side down for a good solid reason, as you will see. In the chapter that follows, you are also free to apply all you have picked up in the sections which preceded. The text is scattered with 'amygdala' and 'long-term potentiation' as much as with 'molecule' and 'aerodynamic'.

The cases in this chapter have explanations of the sort a regular school science curriculum deals with, but they also reek with indignation, that inanimate objects can kick up such a fuss. Feel free to spot animism, naive science and child-like egocentricity on every page. The Laws of Murphy are built on those three mighty pillars.

Your queue always goes slowest

I doubt if there's anyone in a supermarket anywhere in the country who doesn't believe they are standing in the wrong queue. The fact is, though, your queue doesn't always go slowest. You would find this out easily enough if you randomly threw yourself at queues over a couple of months (with a stopwatch of course, and a clipboard and chart), and plotted out the timings of your queue against the others; the result would be disappointingly dull and average. This is one reason why supermarket queues aren't full of people with stopwatches and clipboards. The other reason is that people go to supermarkets to shop, actually.

Why do we imagine the opposite is true? Why do we feel our queue is the slowest? To begin with, let's face it, all queues are slow queues. (Read about the elasticity of time on p. 25.) We hate them. We feel twitchy just at the thought of queuing. We remember the slow queues we were in, but we have no memory of fast queues we were in (or of slow queues we weren't in). This is bad statistics, but we already know that brains aren't hot on things like accuracy. As we edge closer to the tills our gloomy prediction is ready to be supported by the facts (see 'Bad luck comes in threes', p. 191). We expect to be slowest, and lo, we are!

How to pick a queue? This is the question on everyone's mind as they approach the line of tills. They go into hunter mode, eyes darting from side to side, seeking the softest target. Some will aim for the queue that's shortest. Big mistake; it's shortest because there's someone at the end who is trying to revive the ancient art of haggling, and all the queuers behind have fled. Some will jump at the express lanes ('5 items or less'). Big mistake; the slowest staff are on the express tills. Some will go for the longest, thinking there must be a good reason why it's popular; big mistake, obviously.

For your benefit, here is the formula for a comfortable ride through the tills. Factors are: the number of items, the speed of the assistant, the number of queuers and their psychological profile.

First, assess the psychological profile of each queuer, scoring 1–10 on the basis of

- stubbornness (how likely they are to question the price of anything)

- forgetfulness (likelihood of remembering one more thing they need at the last moment and popping off to find it)

- unreliability (likelihood of having lost their purse at the bottom of their bag, or pocket, or is it still in the car? . . .)

- friendliness (likelihood of being close family friend of cash-out Carey and have to ask her about her niece for hours)

- age

Add the scores for all queuers (S).

Second, add the total number of items in all their baskets (x).

Third, paying takes as long as scanning 30 items, so add total number of queuers (n) x 30 to factor in the paying time.

Last, divide by speed of assistant, given by their age (y).

The full formula for each queue is

$(S + x + 30n)/y$

Do this for all the queues at all the tills, then go and start again because all the queues will have changed by now. After the second time through decide it doesn't really matter, join the nearest queue, be happy.

A slightly rougher guide, which has the merit of being practical, is to add up how many basketsful of items there are in each queue (a full trolley counts as 5 baskets). Add 1 basket per person to account for paying time. Join the shortest queue on this estimate.

Once having joined a queue, don't switch because another seems to be going faster. Murphy's Law says that as soon as you move to a faster queue, it stops dead, while the one you've just left races ahead.

In spite of all your cunning, your queue will probably not be fastest. On the other hand it won't be slowest either: it'll probably be average. Boring, but true.

Motorway traffic queues are a special case: here, a scientific explanation is available. Generally the lanes will move forward in turn. You are of course convinced that other lines move forward faster than yours, and you must be dropping back. But if you take a marker from the vehicle next door, you'll see that it actually stays pretty close to your position in the long term. Why does it appear to be going faster?

As you watch the cars next to you begin to move forward, notice that the line of cars lengthens as the gaps between the cars grow. This means that a line of 20 cars grows to the length of 40 when it moves. It certainly seems like 40 cars to you as it rolls by. Then it stops. Then your own line takes off, but you come to a halt after a very short time. You have actually passed those 20 cars, but because they are now bunched together it was a short journey. In your mind you are comparing the **length** of the other line with the **time** of your own journey, which is a wrong comparison. (Take comfort – the drivers in the other line think you are going faster.)

Many places where queueing has reached epidemic proportions, such as airports, struggle to soothe their passengers. Research by psychologists shows that your frustration mounts when you notice that other queues are going faster than yours. So management makes sure that faster queues, such as those of passengers in transit, are hidden behind specially constructed screens. Also your own little queue is masked so you can't see, in the next room, the giant queue you will join when you've finished with this one.

Hospitals learned long ago that patients hate to sit in a room full of hundreds of people who are all before them in the queue. Nowadays you will be called quite quickly, sent down a couple of corridors and placed in another queue. When you get to the end of that queue you'll be eligible for another one in another corridor. This, say the hospitals, makes you feel better. Not so much better that you wouldn't fancy seeing a doctor eventually, though.

Special rule for dealing with crowds trying to enter a single door (such as a tube train or lift in the rush hour)

In this situation rules of queueing break down (i.e. there is a queue, but everyone believes they are at the front). Research has shown that the crowds to the side of the door will move faster. So place yourself at the sides and you'll be through in no time. Alternatively, place yourself at the back in the middle and complain about people who barge in from the side.

MATHS

Why do we hate maths lessons? We aren't born hating maths. We come into the world fitted with the full kit for working with numbers.

An ingenious experiment demonstrates that even two-week-old babies can count. They will watch a toy being placed behind a curtain. If the curtain is lifted to reveal two toys instead if one, they register surprise. (How does the scientist know? He's put a cunning dummy in the baby's mouth, wired up to a pressure gauge to measure how often the baby sucks on the dummy. The suck rate goes up when the baby spots something unexpected − such as two toys being behind the curtain when he is expecting one.)

Babies also show an understanding of size and proportion. The maths basics are inborn. But the child needs to be trained if it is to get beyond the basics. This was true for our hominid forebears in the savannah, and it's even more urgent in the techno jungle of today. For most of us this training is lacking, so we are ripe for Murphy's Law.

In particular, our emotional reaction is equivocal. We don't like cold science or dead maths, so we lend souls and personalities to them. But to fully understand science we should subdue the amygdala, and realize that the inanimate world is truly inanimate. The following observations reveal what victims we become when we enter the Plastic Age with a Stone Age mind:

There are only four numbers: 1, 2, 3, lots

The baby above can count, yes, but not very far. As it grows older it is able to easily recognize what two things or three things look like. Four things is pretty easy to spot, but above there it all gets a little hazy. When it encounters six objects it is unable to instantly recognize 'six', it has to count them one by one. There is some evidence that without the linguistic crutch of the number system, large numbers are difficult or impossible to count with any accuracy, even for adults. Benjamin Lee Whorf developed the theory that, to an extent, our thinking is dictated by the language available. If the words for six and above don't exist in our vocabulary, then we can't count accurately above six. Modern researchers are developing the theory that we are born with two core systems of counting, which don't depend on language at all; a small number system, for instantly spotting two, three and perhaps four objects, plus a large-number system for judging approximations – which of two collections of objects has more in it. The accuracy of our counting beyond those boundaries, the theory goes, is determined by whether we have the words.

Studies of two Amazon tribes bear this out. The Piraha tribe have words for one and two, few and many. That's it. It is found that their accuracy in matching numbers of objects falls away dramatically if more than three objects are used. The Munduruku tribe, on the other hand, have words for numbers all the way to five. Their counting accuracy declines with more than five objects[1].

We, of course, have many more number words, but Murphy's Laws indicate that we don't always use them. When we look at Mickey Mouse we don't say, 'hold on, he's got a finger missing!' Most cartoon characters only have three fingers on each hand, but we don't notice. When we buy a bottle of wine for £4.99 we don't say, 'that's four pounds and . . . good grief, that's nearly five pounds', we say, 'that's four pounds and . . . a lot' (see p. 62; 'Everything is a penny short of a sensible price').

It is interesting to note that in French, the word for three (*trois*) is similar to the word for lots (*très*) and too many (*trop*).

The floating broom illusion: how many fingers on the wrist?

Winning the lottery happens to everyone else

There are two answers to this. The first is that although we all know somebody who knows somebody whose friend won the lottery this year, we ourselves are still miles away from winning. I may have among my friends 100 lottery punters, who all have 100 contacts, who each know 100 other punters. The total number of punters is 100 x 100 x 100 = 1 million. If all of them buy a lottery ticket each week, that's 52,000,000 chances each year, so, even allowing for overlaps, it's a racing certainty that I'm not far removed from a lottery winner.

The second point is that winning the jackpot will definitely **not** happen to you. (How reckless is that last sentence? Well, there are 14,000,000 people in Britain who buy lottery tickets, and 1,700 have become millionaires through the lottery, so (dividing 14,000,000 by 1,700) if 8,200 people in Britain buy this book I'll get one letter of complaint. I can handle that.)

To appreciate this fully you have to have an understanding of probability and chance, an area where human intuition is its own worst enemy. I often see smokers buy lottery tickets. They are thinking 'The chances of winning are one in 14,000,000 . . . it could be me.' As they puff on their fag are they also thinking 'the chance of getting lung cancer are one in 1,000 . . . it won't be me'? Clearly the hard maths is against them here, but the overwhelming addiction to both wealth and tobacco smoke has swamped all sense of logic.

Because lottery numbers are associated with wealth, status, power and endless holidays, they are infused with a sort of life force. We treat the numbers like exotic animals, imagining that they roam around the lottery system like herds of wildebeest . . . 'If the number 23 hasn't cropped up for a while, surely it's bound to happen this week . . .' A friend had one wall covered with a chart on which she plotted the migration of the numbers through the year, hoping to set a trap one week and bag six winners.

Last word to Fran Lebowitz: 'I've done the calculation, and your chances of winning the lottery are identical whether you play or not.'

Any tool, when dropped, will roll into the least accessible corner of the room

On the way to the corner, any dropped tool will first strike your toes

(Arthur Bloch, *Murphy's Law*, 1977)

OK LADS, LET'S MAKE A BREAK FOR IT!

(see also pp. 108–13 for other examples of animism)

It's true and you know it's true, at the same time you know it can't be true. The tool is just as likely to roll into the middle of the floor. A statistical analysis would show the randomness of the scatter, but we would rather believe in the deviousness of tools and the insatiable appetite of cupboards. When we check our memories, our amygdala has ensured that only bad memories, of scrabbling around under the cupboard, have been stored.

You cannot stop the love affair between pasta sauce and a white shirt

It seems churlish to say, 'just add up the number of times the sauce didn't splash over your shirt', but it's true that successful meals are forgotten. Only the disasters count.

The problem is not helped by the eating technique for spaghetti; cover it with a bright red juice then spin it on the fork like a car wash. In this system there's bound to be some collateral damage. In the land of its origin many spaghetti eaters crouch low over the table and approach the plate as if it were an unexploded mine. The Italians have learned respect for Italian cooking.

Campfire smoke always blows in your face

Because our animist minds believe that smoke has motives and goals, we think it's out to get us. In fact the chaotic

nature of campfire smoke ensures that everybody gets a good lungful. But each unhappy camper thinks the smoke has singled them out for punishment. Their knowledge of maths should tell them that random is as random does, but the sad lad who spends his time dodging from one place to another to get out of the smoke will be convinced it veered in his direction on purpose a minute later.

Grapefruit juice squirts into your eye

The technique of most people when tackling the traditional breakfast half-grapefruit is to push the spoon into the opposite side. Because of this any drops that happen to squirt out will fly in the general direction of your face. Your eyes occupy about 1/400th of the blast zone, so the chance of a drop hitting your eye is pretty low. However, your amygdala won't remember it that way: a drop of citric acid in the eye is a memorable event. The amygdala will add enough fear to the memory for it to say, 'look out, it could happen again. Remember the pain last time?'

Traffic lights are always red when you're in a hurry

There goes that amygdala again. How can anything as lifeless as traffic lights conspire against you? Well, I suppose amygdala has a point this time – traffic lights in the centre of cities are usually coordinated across a number of junctions. The traffic managers allow for cars to be travelling at 15 mph, so provided you keep your speed down the lights will change magically in your favour. (I have myself driven along London's Euston Road, with its 14 sets of traffic lights, at a steady 15 mph, with all the lights going green as I arrived at them.) But when you're in a hurry you will be forever sprinting, then braking at the next red lights, then sprinting off again, then screeching to a halt . . . In the end your average speed through the junctions is, of course, 15 mph. But your brain is full of traffic light conspiracy theories.

It is generally true, also, that you are more likely to meet a red light than a green, by a factor of two to one. Modern junctions often involve special filter lights and pauses for pedestrians, so traffic lights usually end up being red about two thirds of the time.

Lifts never stop on your floor

The most irritating thing about using a lift is having to just stand there staring at a blank lift door while you wait for the thing to turn up. Your difficulties with measuring time (p. 25) mean that a 30-second wait with nothing to do feels like an hour. During that age you seethe, psycho-analyse the lift, think increasingly vicious thoughts about its parentage and invent a plot between it, the architect and everyone else in the building to keep it from you. Be patient: while it's being cruel to you it's being kind to someone else on another floor.

When it finally arrives, lets you in, then whisks you off in completely the wrong direction, then you can cuss it good and proper. What you have encountered is the simplest, cheapest lift system, which travels upwards to the top, pick-ing up passengers as it goes, then downwards again, ensuring that at any given moment half the passengers are going the wrong way.

Lift programmers like to make people happy. They really do. But they have a lot of people to satisfy, so they try to spread that happiness evenly among all the passengers. If there are more people calling for lifts at the ground floor than any other – the norm in most buildings – then they arrange for lifts to be biased to go that way, which will make people on other floors cross. In tall buildings with many lifts, lift brains get bigger. They might perform 'zoning', with one lift hovering around the top while another sits at the bottom, and others act as 'midfield sweepers'. Whatever strange antics they get up to, you should remember – the more lifts, the bigger the brains: they know better than you what's good for you.

Perhaps, instead of staring at the lift door while you wait, you could try to work out what programme it's running on. Then time will fly.

Weather forecasts are wrong, wrong, wrong

Weather prediction is where chaos theory makes its nest. Little breezes can suddenly swell up into major storms, then just as suddenly fade away to a light haze, seemingly for no reason. As the wind and clouds leave the Atlantic and start to head across Britain and Europe they are increasingly difficult to predict. Chaos theory tries to explain the inexplicable. In its purest form it expresses that all the weather systems of the world are interconnected, and they're all unstable, so a slight change in one area of the globe can have an knock-on effect way over the other side – a butterfly flapping its wings in Patagonia could theoretically set up a chain of events that causes a hurricane in the Himalayas.

In the weather bureaux they are dealing with a number of factors which will give each mass of circulating air a special character; how big it is, how fast it's travelling, its temperature and humidity. As each mass collides with its neighbour the equations change and things become increasingly wild and woolly. Forecasters use chaos theory to create no less than 50 possible outcomes, then apply a statistical analysis to work out the most plausible story for broadcast. But they very rarely commit themselves to certainty. Because of the selective way you listen (p. 16) you will hear key words – 'rain . . . sun . . . cloud . . . wind . . .' – but if you listen carefully between the key words you will notice the real message; 'might . . . possibly . . . a chance of . . .' In other words, 'your guess is a good as mine, guv'.

(Statisticians will tell you that the most likely prediction for tomorrow's weather is 'the same as today's'.)

Bad luck comes in threes

The maths of bad luck is pretty simple: bad luck doesn't come in threes, it comes in a steady drizzle. But the psychology of bad luck is more complex, as Unlucky Jim finds out below. Highly charged memories can linger for weeks, and link up with others in ways they should not.

The culprit here is **long-term potentiation** (see p. 47), the basic brain operating system. Each of Jim's neurons joins its neighbour at a synapse. When a nerve message reaches the synapse it jumps across to the next neuron and the message continues onward. From then, for a little while, the neuron is primed, and will fire off again given the slightest hint of a repeat signal. Memories are made of neurons linked to each other by this long-term potentiation. For instance, the word 'banana' is stored as a pattern of neurons in the temporal lobe, and this is linked to a site in the visual cortex where there are images of long thin yellow objects. When you hear the word 'banana' the link will produce a visual picture of one.

When the first bit of bad luck happens – Jim slips and breaks his ankle – a memory is created which includes a picture of the banana skin that caused it, the crunch of breaking bones, the pain, the smell of the hospital, and so on. The whole tapestry is electrified by the amygdala's emotional charge; this is **bad news**, it's depressing, it's humiliating. Because of the amygdala's powerful influence, this particular neural tapestry will remain strong for several weeks.

The next day Jim loses his door key. The new memory would normally fade quickly, but some strands – the bad news strands – overlap with the last, still strong, memory of the banana prat-fall. This gives the new memory extra staying power, so two days later, when his grandmother's favourite vase slips through his fingers, the new 'bad luck' memory has two stable-mates. And those three memories now trigger off another memory – that bad luck comes in threes. The belief that bad luck comes in threes becomes a self-fulfilling prophecy. Because Jim believes the fable, then each time there's one disaster his mind will search diligently for disasters number two and three.

Buses come in threes

You didn't get it wrong this time; they really do go round in threes, but they aren't doing it out of spite, nor is it incompetence at the garage: the buses leave the depot at precisely timed intervals. The first bus picks up a load of passengers at the first stop. This takes time. Meanwhile the second bus is getting nearer. By the time the first bus sets off, the interval between them is already reduced. Because of this there is less time for new passengers to arrive at the stop, and because of this the second bus takes less time to pick them up and sets off quicker than number one did. The same happens again at the second stop, and again at the third, with the gap decreasing each time, until the two buses end up running together. On long, busy routes these two can be joined by a third, and then their reputation is secured.

A useful rule of thumb when travelling on public transport is, 'always take the second one'. This certainly works on the London Underground; even at the height of the rush hour, if you wait for number two you can often find yourself on a half-empty train.

Where you want to go is on the edge of the road map

True believers in Murphy's law will know as they reach for the map book that their destination is certain to be on the edge. But what does 'On the edge' actually mean? Usually within 2 cm of the edge, because then it involves lots of irritating page-flipping to check the route. A common page size is 12.5 x 19 cm, i.e. 237.5 cm^2. A 2 cm margin all round adds up to 120 cm^2, which is over half the total area of the page. So you really are likely to feel road-map rage. This is counter-intuitive, but that is the problem with intuitions.

Open-plan maps can't be refolded without using scissors

I asked my maths teacher how many ways there are to fold a map. As always, his answer was (a) preposterous, and (b) accurate.

Here is an example of a pamphlet, to be folded in one direction only, with 4 folds. The grid consists of 5 rectangles of size 5 x 1.

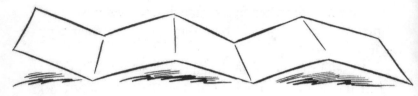

Simple fold.

It will be fully folded when it occupies a space 5 x 1, and has 5 thicknesses of paper. There are $4! \times 2^4$ ways of doing this. 4! means $4 \times 3 \times 2 \times 1$ and 2^4 is $2 \times 2 \times 2 \times 2$, so that works out to $(4 \times 3 \times 2 \times 1) \times (2 \times 2 \times 2 \times 2) = 384$

With n folds, there are $n!$ $\times 2^n$ ways of achieving a complete fold. That means with only 2 folds there are 8 possibilities. If it has 8 panels it can be arranged in 645,120 ways. If it is a map with

8 panels in one direction and 4 in the other, and it is refolded in one direction, then the other, then the total number of folds is the product of the two directions: 645,120 x 48, or 30,965,760. Statistically speaking, I have a better chance of winning the lottery than refolding my map correctly. This fits well with experience.

PHYSICS

Everything ends up in the duvet cover

Here is a section devoted to the quest for the quiet life. The reason why our clothes end up in the duvet cover is the same as the reason why people go to live in the country, also why the Sargasso Sea is where it is, also why street litter goes where it goes.

First the washing. As the socks tumble around in the washing machine they occasionally come across the big heavy duvet lolloping around at the bottom, with its mouth wafting open and shut. If they stray into the mouth, they'll find the turbulence is less there than outside – in other words they move less, so they are more inclined to stay. For every item in the machine, the likelihood of being wafted into the duvet cover is high, the likelihood of being wafted out again is lower. Each new arrival pushes the last one back a bit. The duvet cover fills up steadily.

The same pattern can be seen with street litter: bits of broken glass, grit and pebbles are bounced around by the wheels of cars until they bounce into areas where the car wheels don't go. This is the kerb-side, where I pedal my bike.

In the middle of the mighty currents of the Atlantic Ocean is an area known as the Sargasso Sea. By an accident of geography no winds blow here and the ocean itself is completely still. All the rubbish and weeds being swirled around the ocean have been wafted here and, because there are no currents to waft it out again, here they remain, becalmed. The Sargasso Sea is so clogged with muck that it is by now practically solid.

When we choose our lifestyle we opt for the amount of turbulence we feel we can tolerate. When the hurly-burly of city life becomes too much we bounce out into a nice quiet village. In a strange way, we are behaving like the clothes in a washing machine.

When you're wearing boots, socks that aren't stapled on come off

The heels of the socks are rubbed up and down against the boot. When they are rubbed up they stretch and are pulled down. When they are rubbed down they stay there.

The best skimming stones are farthest from the water

The accepted explanation is that somebody got there before you and skimmed the ones that were near the water's edge. Not so. The flatness which makes them aerodynamic also makes them hydrodynamic. The flat ones are wafted upwards by the incoming tide, while the round ones stay put on the sea bed. When the tide goes out, your favourite stones are left stranded at the high tide line.

There's always one teaspoon left in the washing-up bowl

Teaspoons are different from other items in the washing-up bowl. Small and light, they don't stay put when the water is swirled around the bowl, but swim on the bow-wave of the searching hands. You think you've checked everywhere, but you haven't checked one inch in front of your fingers.

Cornflakes don't fit back in the box

The usual reason for emptying the cornflakes out of the box is to get to the plastic knick-knack inside. When you put the cereal back you find the flakes have grown and no longer fit.

What has actually grown is the air spaces between the flakes. Shaking the box will settle the flakes back together.

Generally you don't want to shake cereal boxes too much because . . .

. . . The dust sinks to the bottom

This is particularly true of muesli, where there is a big difference in the relative size of the pieces; when you peer into a new packet you see nuts and fruit aplenty, but by the time you get to the bottom there's nothing but powder for breakfast. You may think this a trick played by the manufacturer – a bit of window dressing on top of a box of dross. Some try shaking the box so that the pieces are more evenly distributed, but this only makes it worse.

A distant memory of a science lesson on floating and sinking might lead us to believe that larger, heavier pieces should sink, while smaller, lighter pieces should float, but this is not a liquid, and the answer is actually very simple. Larger pieces have larger gaps around them. Smaller pieces drop down into these gaps, so the tendency is for the dust and small flakes to sink down, leaving the nuggets exposed at the surface. Turning the box upside down and shaking will reverse the trend and help spread the dust and the nuggets more evenly.

Hot plates look exactly the same as cold plates

When you or I are hot we show it by looking red-faced and sweaty. Plates are more discreet: they signal that they're hot by jumping out of your fingers and on to the floor.

The automatic reflex that causes you to fling the plate in the air just after you picked it up is a special survival mechanism. If your fingers are being burned, you have to react quickly. Messages from the fingers' pain receptors travel as far as the spine, then straight back to the arm to make it whip the hand out of trouble. Then and only then is the brain allowed to get in on the act, too late to do anything except think 'ouch', apologize to everyone and start looking for a floor cloth. The mind will remember the incident back-to-front (see p. 17). It will say 'I picked up the plate, realized it was hot and dropped it.' The correct sequence of events is 'I picked up the plate, then dropped it, then realized it was hot'.

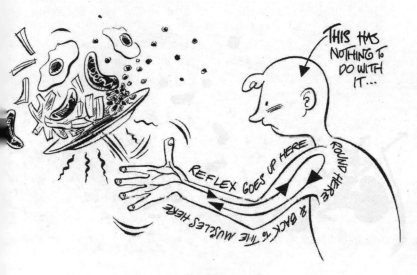

Stubbed toes take their time

That moment when time stands still; you look down at the table leg, and your toe, which has just cracked into it; you think, 'this is going to hurt'; you hear the distant whistle of the oncoming pain. Why does it take so long to arrive?

When the toe met the table leg, three different nerve systems moved into operation. The reflex that jerked your foot away involved large nerve fibres that can conduct impulses at speeds as high as 100 metres (109 yards) per second. These travelled to sites in the spine, then straight back to muscles in the leg, which jerked the foot away from trouble. The second set of nerves sent messages to the brain at about 76 metres (83 yards) per second, informing you there was a problem. But the full accident report took longer. Processing at a stately 1 metre per second, the noisy cavalcade of pain arrived a couple of seconds later, and didn't leave till several minutes afterwards, having partied through your brain, full of noise and colour.

Used batteries are attracted to new batteries

Electricity is a very new thing. It started being studied by scientists as recently as the sixth century BC (which, if you remember that 'modern' man has been around for over a million years – batteries not included – is very recently). In the eighteenth century they thought electricity was some kind of magic liquid. As if to prove this they carried it around in 'Leyden jars' and seemed to be able to pour it out. Today we are no more advanced. We imagine modern batteries are cousin to the Leyden jar, and that electrons pour out of them until they are empty. This leaves us thinking they should look different when dead. They should at least feel lighter. Perhaps this is why we are so cavalier at battery-changing time. Some of us glibly mix the dead up with the live, some of us carefully separate them into two piles, then forget which pile is which.

The perforations in the toilet roll don't match

It doesn't seem that much can go wrong with toilet paper. But sure enough, whatever can go wrong, does go wrong. How can the perforations fail to match up? The answer is, they do match. To restore peace in the loo, take one of the pieces of laminate and unwrap it from the roll, leaving the other behind. Now the pieces match up perfectly.

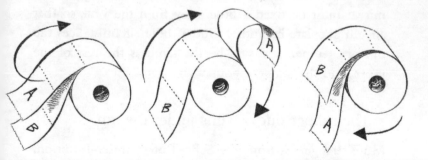

Some supermarkets have had so many complaints from irate customers who think their perforation machines are out to get them, that they've taken to printing operating instructions on the loo-roll wrapper.

The bathroom mirror steams up exactly when you need it

Water floats in the air as individual molecules (water vapour). There's always some there, though you can't see them because the molecules are spread out, and they're moving pretty fast, so when they collide they bounce off each other and go their separate ways. As the hot water pours out of the taps the air becomes full of water molecules. If they meet something cooler – like the bathroom mirror – they are slowed down. Now if two slower molecules collide they won't bounce away, they'll stick together. (Why do they stick together? See p. 206.) When a few trillion water molecules have stuck together they become visible to the naked eye as a droplet. The mist on the mirror is made of thousands of these droplets.

It is possible to have a mirror which doesn't mist up when the bathroom is being used. For this to happen the mirror must be fixed a little away from the wall, so that air can circulate behind it. Under these circumstances the mirror's temperature will be the same as the rest of the room, and the water vapour won't condense.

The rubber duck swims under the tap

More strangeness from the School of Counter-Intuition. As the bath fills up, things floating near the tap should be pushed away, instead of which they are dragged repeatedly under the tap. The phenomenon, known as the Bernoulli effect after the great Swiss scientist who studied it, explains this, and also why chimneys work, why aeroplanes fly, why bingo balls hover on jets of air, why free kicks bend into the net, and why bowlers and tennis players spin the ball.

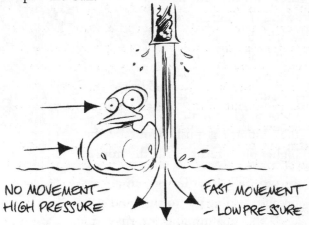

NO MOVEMENT—
HIGH PRESSURE

FAST MOVEMENT
— LOW PRESSURE

- First, **the duck:** when stuff moves it spreads out. By 'stuff' I mean everything: cars spread out on the motorway when they go fast, and bunch up again when they meet speed restrictions. Cornflakes are farther apart as they pour out of the packet than they are in the bowl. Water molecules are farther apart as they pour out of the tap, and for a while after they enter the bath, until they slow down just under the water's surface. So the water molecules rushing past one of the duck's flanks are farther apart than the ones on the other side, so there are fewer of them, so their pressure against the duck's flank is less. The water on the other side is pushing more; the duck heads towards the stream instead of away from it.

- **Chimneys:** when air moves it spreads out too. Moving air is at lower pressure than still air. (When the weather forecaster talks of 'areas of low pressure' they usually mean 'wind'.) Chimneys work because the wind blowing over the top is at lower pressure than the air indoors. The air in the chimney is pushed out by the higher pressure air inside the house.

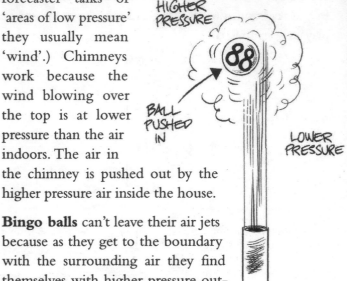

- **Bingo balls** can't leave their air jets because as they get to the boundary with the surrounding air they find themselves with higher pressure outside pushing them back in.

- As **aeroplanes** slice through the sky, the air molecules that travel over the curved top of the wings have to go farther than the molecules below. That simple fact means that the pressure on the top is less than the pressure beneath; the wing is pushed upwards.

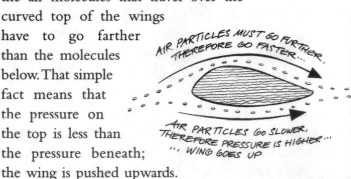

- **Balls** curve when they are spun because as they travel towards their target, with the air rushing past them, one side is moving in the same direction as the air (A), the other in the opposite direction (B). On side A the air is being encouraged to move faster, on side B it is being resisted. On side A therefore the pressure is lower, on side B it is greater, and the ball is pushed in the direction of A.

(A) BOTH FLOW TOGETHER... AIR MOVES FASTER... PRESSURE IS LOWER

(B) BOTH FLOWS AGAINST... AIR MOVES SLOWER... PRESSURE IS HIGHER... PUSHES BALL LEFT

FANTASTIC! I'VE ALWAYS WANTED TO RIDE THE THERMALS!

eeek!

Shower curtains are out to get you

There's nothing clammier than the grip of a plastic curtain as you try to shower in the bath; and there's nothing more certain. Why?

Two reasons: firstly, the water in the shower heats the air, which rises out of the top. Secondly, the rapidly moving water and air reduces the air pressure, so the air outside the curtain moves inward to fill the space, taking the curtain with it.

Why does it cling so tightly once it's found you? See below.

When you most want to pour something carefully, you spill it

It happens all the time: you want to pour just a tiny drop from a glass; the drink gets to the rim and then completely forgets its physics and toddles down the side of the glass, along your arm and off your elbow. What causes water to lose the plot like this? Has it forgotten which way is down?

The answer to this also explains why water droplets form, and why the shower curtain clings. Water is sticky. A water molecule is slightly electrostatically charged, slightly negative at the top and slightly positive at the sides, so when two molecules come close together they tend to be drawn together, like pieces of paper towards a plastic comb.

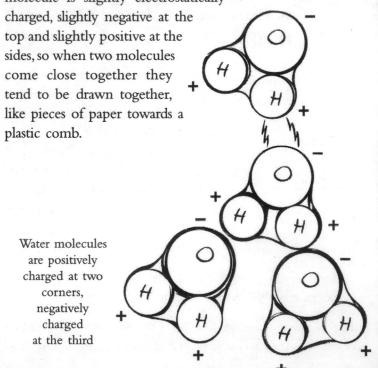

Water molecules are positively charged at two corners, negatively charged at the third

(That's why water is a liquid. It should be a gas, really; a molecule of water weighs slightly less than a molecule of carbon dioxide. The clinginess of the molecules is what keeps it together as a liquid.)

Water molecules don't only stick to each other, they'll stick to anything else they come in contact with, including you (and the shower curtain, see above) and the glass you are pouring from. If you pour very gently the electrostatic attraction of the side of the glass will be more powerful than the force of gravity.

The end of the sticky tape is never there

If you're looking for proof that the world is out to get you, look no further than a roll of sticky tape. As soon as you take your eyes off the end it disappears; you spend the next hour scratching and swearing at the roll. What happens when you're not looking is quite startling. I recommend you try it some time when you don't have anything urgent on. Pull the end a little way off it and watch it being sucked back on to the roll.

The tape has become electrostatically charged. As you pull the end away, it leaves some electrons behind on the roll, with the result that the end is positively charged and the roll, with an excess of electrons, is negative. The two will attract. This also explains why long strips of tape wrap themselves around your arm – they are attracted to you electrostatically.

Toast always lands butter side down

The aura surrounding this, the most famous of Murphy's Laws, is due to the logic-defying truth of it. If you drop a coin to the floor 100 times it will land 50 times heads, 50 times tails. Yet when the toast hits the floor it is with its butter side down 100 per cent of the time. How can this be, except that the World Is Out To Get You?

The BBC did an experiment in 1983 to test the truth of the law: 20 volunteers buttered toast then flung the pieces in the air. Unsurprisingly, the toast landed butter side down 50 per cent of the time. Was the law invalidated? No way. The BBC had got it a wee bit wrong, because we don't often fling toast in the air, even during our wildest, wackiest breakfasts. What does occur frequently is that the toast slips off the plate as you carry it, and as it does so something happens which has nothing to do with Murphy's Law, but a lot to do with Newton's. The toast tilts over the edge, so that as it starts its fall it is rotating. Newton's first law says that when something is doing something it carries on until something makes it stop. The toast will continue rotating until something stops it, in this case the floor. The grief is that only half a rotation is possible before the floor arrives, so the butter side is always the one which hits the dirt.

The solution to the toast problem is simple; always make toast wearing stilts.

GETTING IT IN THE TOASTER IS THE PROBLEM...

The toast will be able to make a full turn before it hits ground zero.

Bugs always drown on your side of the glass

If you believe truly in the malice borne towards you by All Creation (the basis of Murphy's Law), you can have faith that the bug has been stalking you for days, hell-bent on spoiling your picnic. And you might be right. More likely, though, it smelt your drink. Insects have incredibly sensitive smell receptors in their antennae. Your bug may have come several hundred metres to share your drink. Having come all that way it wasn't going to waste time – it went straight for the main course, little realizing that it was a liquid lunch, and thus, for the bug, the last supper.

They don't always drown on your side of the glass, of course, but if you try rotating the glass so the bug moves to the other side, it **stays on your side.** Why so?

When you rotate the glass, that's all you rotate. Newton's First Law says that things stay put unless made to move. The drink isn't stuck to the glass, so it stays in the same position while the glass rotates round it.

Inertia like this lies behind the famous egg trick: you have two eggs, one raw and the other boiled. How can you tell which is

HUH!
YOU SAID
IT'D BE A
WHIRLPOOL
BATH!

which? Just spin the eggs. The raw egg will stop spinning quickly, because all that's turning is the shell, while the contents stay put. Spin a boiled egg, which is solid through-out, and it will turn for longer.

The swimming pool is deeper than you thought

An example of rewriting the back-story (p. 75) is that we can so easily forget this effect. I can look in the pool now without saying 'hey, that looks a heap shallower than it should', but there was a moment in the past when this illusion nearly did for me. I was seven and this was my first go in a grown-up's pool. As I looked into the water I decided it was just about shallow enough; it would come up to my chin. I lowered my feet gingerly to where my eyes told me the floor was, with my mouth just out of the water. When I let go of the side I sank another foot, up over my ears. This was the first time I realized that there were untamed forces at work out there.

The speed of light is not constant. It is slowed down by whatever substance it passes through, the denser the substance, the more it is slowed. (So its speed ranges, as one wit put it, from 299,792,458 metres (over 186,000 miles) per second in a vacuum to 0 metres per second in a

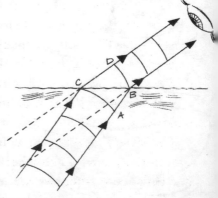

brick.) The drawings show a
beam of light from the floor
of the pool leaving the water.
As the light escapes it speeds
up, so C will move to D in
the time A takes to get to B.
Consequently the whole
beam bends (it is refracted).
Your eye will see it coming
from a different place. When
that happens to both eyes,
your brain sees the floor in the wrong place.

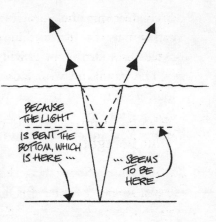

BECAUSE THE LIGHT IS BENT THE BOTTOM, WHICH IS HERE ...

... SEEMS TO BE HERE

Once you get on a bike, all roads are uphill, the wind is against you, the rain is in your face

This illusion highlights the egocentrism which we are all
born with, and which we never completely grow out of.
When the pedalling gets harder we know that the hill is
becoming steeper. We should be thinking, of course, that
we are becoming more tired, but we don't; 'man is the
measure of all things', said Protagoras, but 'man' is an unreliable ruler. We assume our strength to be a constant and
measure the world against that. When swimming, we notice
that water becomes thicker as it is stirred by our arms. Is
it really becoming denser, or are we too dense to understand what's going on?

The 'wind is against you' illusion is another example of
egocentricity, one which most of us will remember from
our childhood; when we set off in a car on a perfectly
windless day, a breeze appears from nowhere. The breeze

is created, of course, not by the air rushing through the car, but by the car rushing through the air. When you cycle through the air it will feel like a wind in the face.

The 'rain is in your face' phenomenon has a similarly logical solution. If the rain is falling vertically and you are cycling forwards then, relative to you, it is falling backwards. You are cycling into it, so of course it will land on your face.

BIOLOGY

Insects spend all day trying to get out of the house and all night trying to get in again

Insects are just as much victims of Murphy's Law as we are. The complex world we have created over the last 4,000 years has left them confused. They haven't evolved an understanding of glass, so they continue to be devoted to their old ways of escaping from somewhere that's dark, to head towards the light. For 200 million years that was a good system, but now it's a bit of a headache for them. It is entirely possible that in time flies will evolve a way out of the problem. Just as hedgehogs are evolving to react to approaching headlights by running rather than by rolling up in a ball, flies may evolve a mental algorithm which will allow them to leave houses by the same route that they came in by. But for the time being flies are going to stick to head-butting windows.

Are we that much more advanced than a fly? Most of us have walked at some time into a glass door that we haven't spotted, so we can sympathize a little. On the other hand, once having bumped we don't keep going back for more.

Why do insects all try to come indoors at night? Many romantic allegories refer to moths diving suicidally into a candle, hypnotized by its brightness. The evidence suggests that this is a myth – a classic example of post-Palaeolithic thinking: we see what we want to see, we build a rich and plausible science to explain it, we ignore contradictory evidence. There's nothing to stop you doing the experiment. You don't need permission from the Science Research Council, or a government grant. Just light a candle in your room and leave the windows open. Insects will fly in, but they won't dive into the flame, they'll sit on the walls around the room, like a crowd waiting for the cabaret to begin. And there are as many theories about why this happens as there are species in the room. Some scientists believe that the creatures mistake the light for the moon, and try to use it to navigate by. Some believe that, just like daytime insects, they associate brighter areas with the sky and freedom. Then again, perhaps they settle on the walls around because they think it's daytime, and fall sleep. Or perhaps they are dazzled by the brightness, and can't fly away from it because everywhere else seems impossibly dark. One day moth psychologists may be able to make fMRIs so small that they can see how the moth's brain is working. Until then, we must pick between the moth myths.

Slugs go straight for the lettuce

The sense of personal outrage that we feel when slugs single out our favourite lettuces from all the possible plants there are in the garden. They're only slugs, after all; why can't they be happy with dock leaves? How dare they act so pretentious!

Slugs prefer lettuce leaves to dock leaves for exactly the same reason we do – they are less poisonous. Over the past 200,000,000 years plants have created a bewildering array of chemicals in their sap to protect themselves from being eaten. They include inorganic and organic salts, toxic minerals, alcohols, aldehydes, alkaloids, polyphenolic compounds and salicylates.

Some of these poisons we have adapted for our own use, as semi-poisons – tobacco, quinine, caffeine and cocaine among many others. Some have found medicinal uses, such as aspirin, found in the bark of the willow tree; or digitalis, a heart stimulant (or poison if you use too much) derived from the foxglove. Some have been carefully bred out of the plant stock – potatoes and tomatoes are both descended from deadly nightshade (and you can see how closely related they are if you eat green potatoes, which are poisonous).

Lettuces are from the sunflower family. They have been cultivated for at least 2,500 years to be as sweet and poison-free as possible, that's why you, I and the slugs all race for them.

The more it's good for you, the worse it tastes

(See also '"Good for you" means "yucky"', p. 13 and 'Whatever you like is illegal, immoral or fattening', p. 79).

The trouble with innocuous food is that it has less of what we need. Some of those poisons I mentioned above really are quite useful. Vitamins, beta-carotene, folates, calcium and potassium are all found in greater quantities in darker vegetables, with stronger tastes.

Fashions in food change little, and take a long time to do so, because the taste and smell senses are so closely connected to the amygdala. A feeling of disgust is never far away (see p. 13). Anything our taste buds classify as 'different' is approached warily. This wariness is a survival mechanism: what we eat could poison us, and once inside us we won't be able to do much about it.

Children are born with certain inbuilt criteria and their simple tastes in food reflect that. They'll turn their noses up at many of the things grown-ups take for granted: curry, coffee, beer, mustard, Stilton, pepper or pickled onions. But they'll happily accept fruit, pasta, milk, butter and bread. The smell of rancid milk produces an instinctive jerk of the head in disgust, but if introduced in the right way there a couple of kinds of mouldy milk that children will take to – yoghurt and cheese. Acquiring the taste for more exotic food is a social act – they need the reassuring company of people who are drinking coffee but not falling dead, before they'll attempt anything so reckless themselves.

Unborn babies are even pickier. When a woman is pregnant, the embryo eats whatever she eats, so she will tend to be careful what she takes in. Mothers will often feel nauseous at the smell of certain foods. Many doctors think morning sickness is an ultra-careful reaction, to protect the fetus from the slightest hint of harm from poisons in food.

Either the hot pool is too hot or the cold pool is too cold. Or both

Leisure centres often have two or more pools, with different temperatures. These pools will exercise your homeostasis system. The body operates comfortably within a very narrow temperature range (20–24 °C or 68–75 °F). The hypothalamus, in the centre of the brain, acts to keep you within that range. Become too hot and capillaries will be enlarged at the skin surface, routing blood near to the surface to help heat loss (you'll show this by going red), sweating will start, to remove heat through evaporation, and if the heat continues you will become lethargic, which the body's way of discouraging you from exercising and getting even hotter.

Once you have adapted this way to the cosy hot pool, the cold pool will seem freezing. The hypothalamus will get down to work reducing capillary blood-flow (you'll go pale), stopping the sweat and, if you don't start to splash about energetically, getting you to shiver to create heat. Once your temperature is back up again and your homeostasis mechanisms are ready to relax a little, a return to the hot pool will shock you with how much hotter it has become.

You are made to get into the sea when it's cold; you are made to get out again once it has become nice and warm

It's all very well coping with the artificially controlled temperatures of the leisure centre, but out in the open everything is in the same environment, so shouldn't it all be the same temperature? Shouldn't the sea feel just as warm as the sand? The sun is shining down on both of them equally, after all.

Sunlight doesn't heat the sea or the air up directly; it shines right through them and heats up the ground, which then heats the air and the water by conduction. (This is a little simplified; it does heat water a little as it shines through it, but not much.) The ground warms the air faster than it warms the sea, so even though it feels warm in the sunshine you should stay out of the water till later. (And I don't mean later in the day, but later in the month.)

Well OK then, jump in. After a little while you will adjust (see above) and the water will feel pretty warm. But by then it'll be time to go home. That's Murphy for you.

As soon as you turn on the light inside it goes back to midnight outside

You are getting up just before dawn, when the sky outside is light, but not bright. As soon as the room light goes on, darkness descends outside; it seems that as you turned the light on inside, you turned it off outside.

Your eye adapts to light conditions directly, by adjusting its aperture. When it is dark the iris expands to allow more light into the eye. The dim light of the sky seems quite bright as you survey it with dark-adjusted eyes. As the light goes on in the room, your iris muscle jumps into action, cutting out the glare, but reducing the outside to relative darkness.

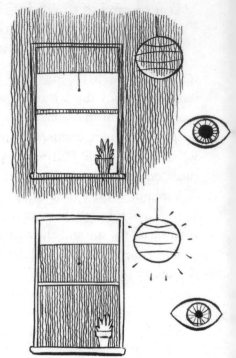

The pupil dilates in the dark, so more light enters from outside – it seems light. When the pupil contracts in the light, less of the external light gets through.

SUMMARY

Ironically, the more pure science explains the world's oddities, the more it drains the life from it. Truth, like Mars, is grey, dull and lifeless. We want it to be red, romantic and populated. Sorry!

Many ancient stories of the world are not only more fun than dry science, but also seem quite plausible. For instance, Aristotle's theory of gravity is brilliant: the Universe is constructed from four elements; earth, air, fire and water. Each has a desire to return to its natural home; fire up in the sky, earth down to the ground, etc. So the apple falls down from the tree because it is going home to Mother Earth, while flames go upwards because they are going home to Heaven. Everything fits. Aristotle's theory was not only completely wrong, it was also the only accepted theory for 2,000 years.

The mighty pillars of naive science, animism and egocentricity held up science, in both senses, for 2,000 years, from Aristotle's time until the Renaissance. We like to think of the ancient ways being swept aside by the bright, clear minds of the Renaissance, the Age of Reason, and the Industrial Revolution; lucid, no-nonsense thinking from Copernicus, Newton and Einstein should have illuminated all the dark corners. But Murphy's Laws show that there are still a few shadows. There's a lot of sweeping and a load of illuminating to be done yet.

9

Fuzzy Logic

Murphy's Laws, when they happen, should remind us how very well we manage to cope the rest of the time, when they don't happen.

After all, we aren't working with the best materials. The human species was cobbled together from bits left over after the last mass extinction. Before that bad day, 65 million years ago, there were many other even worse days which nearly did for life on the planet again and again. And since then there have been more scorchings, freezings, floods and bombardments. Earth and its inhabitants have been used for cosmic target practice throughout its 5 billion year existence, and now, as the dust settles, here stand **we**, not the shining apex of a smooth process of refining perfection, but chance survivors, stumbling from the shooting range. That we're here at all is pure fluke. Were it not for the Chicxulub meteor 65 million years ago, we'd still be nothing more than dinosaur snacks. So don't expect to look at the human machine and find perfection; expect junk; not so much *Homo sapiens* as 'Homo odds-an'ends'.

And what junk there is: junk DNA, 90 per cent of which, many geneticists have told us, is a waste of space; bits of junk body (look at your goosebumps and despair); a junk brain, packed with oddities, compromises and quick fixes. (Who could possibly suggest, for instance, that the corpus callosum is a good idea? The brain throttled down the middle, forced to think in two halves. Wiring trailing all over the place – the left ear with its related cortex on the right side of the brain and vice versa; the visual cortex at the back of the skull, as far from the eyes as you could get.)

The strange ways of Murphy's Laws demonstrates this jumble. Chapter 1 shows we aren't very certain what we are looking at and Chapter 2 shows we're hopeless at measuring it. Chapter 3 shows we have fuzzy memories, and Chapter 4 that we use them to jump to the strangest conclusions. All of which is of no consequence because Chapter 5 shows that we much more readily believe what our emotions tell us, and anyway, as Chapter 6 shows, what we think is largely dictated by what everyone around us thinks we should think. From déjà vu to voting patterns, there is a fuzziness all round.

None of this mattered in the Palaeolithic landscape. In the days when chipped flint was eyed suspiciously as new-fangled technology, the world was very simple, very predictable, so there was no place for Murphy's Law. (That toast couldn't land butter-side down, for instance, until we had invented many things; butter, the knife, the plate and, of course, sliced bread.)

In those simple days Early Man had learned all he needed to know by the time he was ten, and the world served up few surprises from then on. The rocket-propelled days we live today leave us jet-lagged. We do our best to learn the new ways as they arrive, but we're always going to be skidding on the corners.

If the mind we have inherited is not big enough for modern times in some ways, in other ways it is too big. The basic urges – the 'four Fs' of feeding, fighting, flee-ing and fornication – have been put under pressure by

Here's an example of the skidding mind trying to keep a grip. We've taken a few years to get used to the very modern novelty of the animated cartoon. In the very early days Walt Disney discovered that getting the lip sync right on the characters in Snow White was not easy. If the mouth animation was perfectly synched to the words, the audience perceived them as behind the words. Their brains, it seems, were struggling with this newfangled animation stuff, and that skewed their sense of timing. Walt Disney spotted the problem and put the edit slightly out of sync, placing the sound of the words slightly ahead of the picture. This helped the audience, everyone was happy and the films seemed in sync, even though they were now very slightly out. Modern audiences have learned the cartoon meme, their brains work faster, and early Disney classics now seem out of sync the other way.

civilization: the urge to eat has not proved easy to control in a world full of tempting tastes; road rage reignites our instinctive fight/flight reflexes; and as for sex, it is still very hard to say, 'OK, that's enough children, let's stop doing sex now', and switch it off.

Of particular relevance to Murphy's Laws is the difficulty of controlling our mirror neurons. That ability to empathize with our surrounding group tends to overspill on to the surroundings – we empathize with inanimate objects, and therefore feel free to resent them when they won't behave.

So this is the beast that is let loose on the planet; it thinks as little as possible, it carries a knapsack of expectations which on the whole it prefers to reality, it jumps to conclusions based on little or no evidence and it thinks objects have souls. Would you trust a mind like that with your life? Would you trust it with a wooden jigsaw, even?

If you are by now attuned to Murphy's Laws you may find that they pop out at you a lot more than they did before as you go about your daily business. You can do it deliberately: persuade yourself that things that occur together are caused by each other and it will become so; just for the exercise, tell yourself that people a little way away from you are indeed small people, or that the writing on this page is not flat but lumpy, or that a politician you really trust (if you know one) is secretly plotting the downfall of your government, and you will be surprised how easy it is for reality to fall in line with the new order.

And yet the wonder is that we do manage to make sense of nine tenths of the planet nine tenths of the time. For that we should be grateful. Murphy's Laws are only a tiny part of our life. If looked at positively they could be a major part of understanding how we live.

Appendix I

ALL MURPHY'S LAWS

UNIVERSAL

HOME

As soon as you turn on the light inside it
goes back to midnight outside 217

SELF

Conscience: the inner voice that warns us
someone might be looking 15

When your hands are tied your nose itches 20

For any given buffet there is one too many plates 18

The longer you look at the page, the more
the words don't go in 16

Déjà vu has to be seen to be believed 56

You think of 10 important things to remember
just as you are falling asleep 20

Standing on a cliff edge is the one time you
need to be cool, and the one time you can't 9

Bad luck comes in threes 191

Nostalgia ain't what it used to be 77

Life can only be understood backwards, and
can only be lived forwards 75

In youth the days are short and the years are long;
in old age the years are short and the days long 27

The words a man can never utter and live:
'might you possibly be a little premenstrual?' 100

If parents shout at you, it's because they're wrong.
If they don't shout at you, it's because you're
right 102–3

Her pile is mess; your pile is work in progress 153

Hire a teenager, while they know everything 103

The music is too soft for you and too loud for them 39

How long a minute is depends on which side of
the bathroom door you are on 26

SOCIETY

WORK

PASSION

BEASTLY THINGS

Appendix 2

THE fMRI SCAN

The functional magnetic resonance imager (fMRI) is a great boon to understanding how the brain works. It is always hard to analyse something when it is functioning smoothly, so the history of brain research mostly involved studying brains that had gone wrong. This involved a continuous search for damaged brains that could be analysed, preferably while the owners were still alive. Battlefields were a rich source of damaged brains for study in the early days. Later on, industrial accidents provided more. Recently the motor car has furnished many specimens. It's all, frankly, gruesome. Sitting around the emergency department of a hospital waiting for not-quite-dead people to be wheeled in is not an occupation for respectable folk.

More seriously, a person in such a traumatized state is in no position to behave like a normal person would. They've got big problems, and they don't get any better while the neurosurgeon is tinkering inside their heads. 'You're never the same once the air hits your brain' is the neurosurgeon's motto.

The fMRI has changed all that. It is able to study changes in normal brain processes as they happen, even when the changes in thinking patterns are tiny, and all without any invasion of the skull at all. Something as small as the decision to move a finger shows up on the fMRI screen. At last it is possible to watch what happens when nothing much happens.

As the name implies, magnetic resonance imagers work with magnetism. When an area of the brain is active it needs oxygen. It pulls in red (oxygenated) blood and pumps out blue (deoxygenated) blood. Here's the trick; blood is slightly magnetic, because it has iron in it. (Have you ever looked at dried blood and thought it looks a little like rust? That's why.) Even more interestingly, blue blood is slightly more magnetic than red. The amount of magnetism is tiny, of course, but if a large enough magnet is used the changes in magnetism can be measured. The fMRI is a circular magnet, 2 metres (over 6 feet) in diameter, with a hole in the middle big enough for the patient to lie in. The magnet is so powerful that if anyone wearing a watch should come within two metres of it, the watch and attached arm will be sucked straight into the hole. Linked to a powerful computer, the fMRI can provide detailed 3D pictures of brain processes.

The fMRI can be used to spot brain tumours at an early stage, by identifying areas which are demanding an unusually large supply of blood, But it is so subtle that it can even register which parts of the brain are active when the patient simply thinks. Many of the findings in this book have only been possible since the development of the fMRI.

References

TAKING IT ALL IN:
THE SENSES

1 Cherry, E.C. (1953) 'Some experiments on the recognition of speech, with one and with two ears'. *Journal of the Acoustical Society of America*, **25**, 975–79.

2 Sacks, O. (1985) *The Man Who Mistook his Wife for a Hat*. London: Duckworth.

3 Wiseman, R., Watt, C., Stevens, P., Greening, E. and O'Keeffe, C. (2003) 'An investigation into alleged "hauntings"'. *British Journal of Psychology*, **94**, 195–211.

4 Eysenck, M.W. and Keane, M.T. (1995) *Cognitive Psychology*. Hove: Lawrence Erlbaum Associates.

GETTING THE MEASURE

1 Cited in 'Time perception', in *Handbook of Experimental Psychology* (ed. S.S. Stevens). New York: Wiley, 1951, p.1231.

2 Sacks, O. (1973) *Awakenings*. London: Duckworth.

3 Bloomer, C.M. (1976) *Principles of Visual Perception*. New York: Van Nostrand Reinhold.

MEMORIES

1 Rose, S. (1992) *The Making of Memory*. London: Bantam.

2 Nisbett, R. and Wilson, T. (1977) 'Telling more than we can know: Verbal reports on mental processes'. *Psychological Review*, **84**, 231–59.

MAKING CONNECTIONS

1 Roethlisberger, F.J. and Dickson, W.J. (1939) *Management and the Worker: An account of a research program conducted by the Western Electric Company, Chicago*. Cambridge, Mass.: Harvard University Press.

2 'The whole truth'. *New Scientist*, 182(2445), 1 May 2004, p. 3.

3 Nader, K. (2003) 'Memory traces unbound'. *Trends in Neuroscience*, **26**(2), 65–72.

EMOTIONS

1 Schachter, S. and Wheeler, L. (1962) 'Epinephrine, chlorpromazine and amusement'. *Journal of Abnormal and Social Psychology*, **65**, 121–8.

2 Bartels, A. and Zeki, S. (2000) 'The neural basis of romantic love'. *NeuroReport*, **17**(11), 3829–383.

3 Dutton, D.G. and Aron, A.P. (1974) 'Some evidence for heightened sexual attraction under conditions of high anxiety'. *Journal of Personality and Social Psychology*, **30**, 510–17.

4 Raine, A., Buchsbaum, M., and LaCasse, L. (1997) 'Brain abnormalities in murderers indicated by positron emission tomography'. *Biological Psychiatry*, **42**, 95–508.

5 Glaser, D. (2000) 'Child abuse and neglect and the brain'. *Journal of Child Psychology and Psychiatry*, **41**(1), 97–116.

6 Horn, M. (2002) 'Premenstrual syndrome as a criminal defense in the American legal system'. *LA 401: Science, Technology, and Human Values*. Baker University.

7 Sowell, E.R., Thompson, P.M., Holmes, C.J., *et al.* (1999) 'In vivo evidence for post-adolescent brain maturation in frontal and striatal regions'. *Nature Neuroscience*, **2**(10), 859–61.

8 Marsicano, G., Wotjak, C.T., Azad, S.C. *et al.* (2002) 'The endogenous cannabinoid system controls extinction of aversive memories'. *Nature*, **418,** 530–4.

9 Heider, F. and Simmel, M. (1944) 'An experimental study of apparent behavior'. *American Journal of Psychology,* **57** (2 April), 243–59.

10 Damasio, A.R. (1994) *Descarte's Error: Emotion, Reason and the Emotional Brain.* New York: Putnam.

PUBLIC OPINION

1 Rizzolatti, G. and Arbib, M.A. (1998) 'Language within our grasp'. *Trends in Neurosciences,* **21**(5),188–94.

2 Gopnik, A., Meltzoff, A. and Kuhl, P. (1999) *The Scientist in the Crib: Minds, Brains, and How Children Learn.* New York: William Morrow & Co.

3 Watanabe, K. (1994) in *The Ethological Roots of Culture* (ed. R.A. Gardner *et al.*), Dordrecht: Kluwer, pp. 81–94.

4 Sherif, M. (1935) 'A study of social factors in perception'. *Archives of Psychology,* **27**, 187.

5 Ekman, P., Davidson, R.J. and Friesen, W.V. (1990) 'Emotional expression and brain physiology II:The Duchenne smile'. *Journal of Personality and Social Psychology,* **58**, 342–53.

6 Asch, S. (1946) 'Forming impressions of personality'. *Journal of Abnormal and Social Psychology,* **41**, 258–290.

7 Adler, R. (2000) 'Pigeonholed'. *New Scientist,* **167**(2258), 38.

8 Bargh, J.A., Chen, M. and Burrows, L. (1996) 'The automaticity of social behaviour: direct effects of trait concept and stereotype activation on action'. *Journal of Personality and Social Psychology,* **71**, 230–44.

9 Goldstein, J.S. (2001) *War and Gender: how gender shapes the war system and vice versa.* Cambridge: Cambridge University Press.

10 Mbokasi, B., Visser, D. and Fourie, L. (2004) 'Management perceptions of competencies essential for middle managers'. *SA Journal of Industrial Psychology,* **30**(1), 1–9.

11 Milgram, S. (1963) 'Behavioural study of obedience'. *Journal of Abnormal and Social Psychology*, **67**, 391–8.

12 Zimbardo, P. (1973) 'The Stanford prison experiment'. *Cognition*, **2**(2), 243–55.

13 Asch, S.E. (1955) 'Opinions and social pressure'. *Scientific American*, **193**(November), 31–5.

14 Festinger, L. (1957) *A Theory of Cognitive Dissonance*. New York: Harper & Row.

15 Schechter, B. (2002) 'Push me, pull me'. *New Scientist*, **175**(2357), 24 August, 45–7.

16 Holmes, B. (2001) 'In search of God'. *New Scientist*, **170**(2287), 21 April, 24.

PURE SCIENCE

1 Butterworth, B. (1999) *The Mathematical Brain*. London: Macmillan.

Further Reading

A small number of books hit the spot for me over and over:

Rita Carter, *Mapping The Mind*. London: Weidenfeld & Nicolson (1998)

Richard Gross, *Psychology*. London: Hodder and Stoughton (2001)

Rob Eastaway and Jeremy Wyndham, *How Long is a Piece of String?* London: Robson Books (2002)

Rob Eastaway and Jeremy Wyndham, *Why Do Buses Come in Threes?* London: Robson Books (1999)

Joseph LeDoux, *The Emotional Brain*. London: Phoenix (2004)

Steven Rose, *The Making of Memory*. London: Vintage (2003)

And of course:

Arthur Bloch *Murphy's Law*. Los Angeles: Price/Stern/Sloan Publishers Inc. (1977)

Index